Vladimir P. Skripov and Mars Z. Faizullin

Crystal-Liquid-Gas Phase Transitions and Thermodynamic Similarity

Related Titles

Baidakov, V. G.

Explosive Boiling of Superheated Cryogenic Liquids

approx. 400 pages with approx. 105 figures and approx. 15 tables
2006
Hardcover
ISBN 3-527-40575-5

Sinaiski, E., Lapiga, E. J.

Separation of Multiphase, Multicomponent Systems

approx. 550 pages with approx. 240 figures and 22 tables
2006
Hardcover
ISBN 3-527-40612-3

Schmelzer, J. W. P.

Nucleation Theory and Applications

472 pages with 157 figures
2005
Hardcover
ISBN 3-527-40469-4

Mazenko, G. F.

Fluctuations, Order, and Defects

540 pages
2003
Hardcover
ISBN 0-471-32840-5

Kostorz, G. (ed.)

Phase Transformations in Materials

724 pages with 392 figures and 33 tables
2001
Hardcover
ISBN 3-527-30256-5

Mazenko, G. F.

Equilibrium Statistical Mechanics

630 pages
2000
Hardcover
ISBN 0-471-32839-1

Senatskommission zur Beurteilung von Stoffen in der Landwirtschaft (ed.)

Transient Phenomena in Multiphase and Multicomponent Systems

Research Report

395 pages with 241 figures and 10 tables
2000
Softcover
ISBN 3-527-27149-X

Vladimir P. Skripov and Mars Z. Faizullin

Crystal-Liquid-Gas Phase Transitions and Thermodynamic Similarity

WILEY-
VCH

WILEY-VCH Verlag GmbH & Co. KGaA

The Authors

Vladimir P. Skripov
Institute of Thermal Physics of the Ural Branch
of the Russian Academy of Sciences
itp@itp.uran.ru

Mars Z. Faizullin
Institute of Thermal Physics of the Ural Branch
of the Russian Academy of Sciences
faizullin@itp.uran.ru

Consultant Editor

Jürn W. P. Schmelzer
University of Rostock
Physics Department
juern-w.schmelzer@uni-rostock.de

Library of Congress Card No.:
applied for

British Library Cataloguing-in-Publication Data
A catalogue record for this book is available from the British Library.

Bibliographic information published by Die Deutsche Bibliothek
Die Deutsche Bibliothek lists this publication in the Deutsche Nationalbibliografie; detailed bibliographic data is available in the Internet at <http://dnb.ddb.de>.

Printing Strauss GmbH, Mörlenbach
Binding J. Schäffer Buchbinderei GmbH, Grünstadt

Printed in the Federal Republic of Germany
Printed on acid-free paper

ISBN-13: 978-3-527-40576-3
ISBN-10: 3-527-40576-3

Contents

Crystal–Liquid–Gas Phase Transitions and Thermodynamic Similarity. Vladimir P. Skripov and Mars Z. Faizullin

ISBN: 3-527-40576-3

Foreword

This monograph is written by two outstanding specialists in the field of the experimental and theoretical analysis of first-order phase transitions, Academician Prof. Vladimir P. Skripov and Dr. Mars Z. Faizullin. It presents for the first time a complete overview on the research of both authors on the comparative analysis of solid–liquid and liquid–vapor phase transitions, their similarities and differences with special emphasis on the aspects of thermodynamic similarity.

Prof. Skripov has been the founder of an outstanding school of research in the above mentioned field; for several decades he worked as the director of the Institute of Thermal Physics of the Ural Branch of the Russian Academy of Sciences in Ekaterinburg, Russia. His results are published in four monographs and more than 300 journal publications. In particular, I would like to mention his monograph *Metastable Liquids* (Nauka, Moscow, 1972), published in the English version by WILEY in 1974, which is still highly popular among scientists dealing with the processes of boiling of liquids. Prof. Skripov is presently apppointed as a Councillor of the Russian Academy of Sciences.

Dr. Faizullin has been engaged in the problems of thermodynamic similarity in liquid–vapor and crystal–liquid phase transitions for more than two decades. He defended both his PhD and DSc theses on related topics. He is the author of more than sixty articles in scientific journals and one monograph. Presently, he is deputy director of the Institute of Thermal Physics of the Russian Academy of Sciences in Ekaterinburg, Russia.

The results of their long-standing highly original investigations have been presented by the authors and discussed with much interest in several of the research workshops *Nucleation Theory and Applications* at the Bogoliubov Laboratory of Theoretical Physics of the Joint Institute for Nuclear Research in Dubna near Moscow organized by the editor of the present book regularly each year since 1997. First accounts of the results are published in the workshop proceedings (Nucleation Theory and Applications, Dubna 1999 and 2002) and in the monograph *Nucleation Theory and Applications* published by WILEY-VCH in 2005. It is a real pleasure to have the opportunity to present now the extended English translation of the monograph of the authors published in Russian in 2003. I believe, the present monograph can be of similar impact on the research in the field of first-order phase transitions as the already cited monograph of Vladimir P. Skripov published by Nauka in 1972 and by WILEY in 1974.

In the title of the book the term "gas" but not "vapor" is used. It is convenient for a designation of the three states of aggreation of matter: crystal, liquid, and gas. But in discussing the coexistence of the different phases it is preferable to use the term "vapor" but not "gas". In such a way, it is done in this book.

Crystal–Liquid–Gas Phase Transitions and Thermodynamic Similarity. Vladimir P. Skripov and Mars Z. Faizullin

ISBN: 3-527-40576-3

Finally, I would like to acknowledge the valuable assistance of Dr. Irina G. Polyakova (St. Petersburg, Russia), Dr. Alexander S. Abyzov and Andrew A. Abyzov (Kharkov, Ukraine) in the preparation of the book for publication.

Rostock (Germany) & Dubna (Russia), August 2005 Jürn W. P. Schmelzer

Preface

In this monograph, a wide spectrum of thermodynamic aspects of first-order phase transitions is analyzed. Hereby the analysis is extended beyond the range of phase coexistence of stable phases in order to incorporate phase coexistence of metastable states and an analysis of the limits of stability of the metastably coexisting phases. This extension of the analysis allowed us to arrive at a variety of previously unknown relationships reflecting the thermodynamic similarity of different one-component substances in the phase transitions crystal–liquid and liquid–vapor. This approach is then extended to the description of phase equilibria of binary solutions with upper and lower critical dissolution points.

As already mentioned, one of the specific characteristics of this monograph consists in the extension of the analysis of liquid–vapor and crystal–liquid phase coexistence to metastable states. In some respects, these two classes of phase transformation processes behave similarly; however, there exist a variety of features distinguishing these transformations. For example, a coexistence of liquid and vapor can occur at positive pressures, exclusively. With an increase of temperature, the phase coexistence is terminated at the critical point where the both coexisting phases become identical. The thermodynamic properties of the fluid at the critical point may serve as scaling parameters allowing us to establish the similarity of different substances with respect to liquid–vapor phase transitions. The critical point is simultaneously characterized by the approach of the boundary of stability of the fluid. In contrast, the melting curves do not exhibit such high-temperature limit of phase coexistence. In addition, liquid–crystal phase equilibria may be preserved also at negative pressures (i.e., if both phases are exposed to some tensile stress). The melting curve has a metastable continuation to lower temperatures beyond the triple point of crystal–liquid–vapor equilibrium coexistence. In the case of crystal–melt phase coexistence, scaling parameters may be chosen established by employing characteristic parameters obtained for the asymptotic limit $T \to 0$. With an increase of the tensile strength (or an increase of the absolute value of the negative pressure to which the system is exposed), the degree of mechanical stability of both liquid and crystal coexisting phases is decreased but the stability boundaries are not necessarily reached.

This monograph is basically a translation of the Russian version of the book published recently (V. P. Skripov, M. Z. Faizullin: *Phase Transitions Crystal–Liquid–Gas and Thermodynamic Similarity*, Fizmatlit, Moscow, 2003 (in Russian)). However, some additional paragraphs are added. In Section 2.3, a discussion of the universal behavior of different substances in the approach of the liquid–vapor critical point is given. This analysis is connected with the experimentally observed crossover from classical mean-field to a nonclassical behavior. In Section 3.5, estimates of the magnitude of the internal pressure in metals are given, in Section 3.6 the limits of thermodynamic stability of the condensed phases of alkali metals are discussed and in Section 3.10 a computation of the change of the entropy at an isothermal homophase expansion of solid metallic substances is performed.

Finally, the authors would like to express their deep gratitude to Prof. J. W. P. Schmelzer, who performed much work in the preparation of the English version of our monograph for publication.

Ekaterinburg (Russia), August 2005 Vladimir P. Skripov Mars Z. Faizullin

1 Introduction

1.1 Basic Aims and Methods

The problems of the thermodynamic and kinetic description of equilibrium phase transitions of first order are discussed in various original publications and manuals. It may seem that the commonly employed general approach to the different forms of phase transitions based on Gibbs's theory is quite sufficient for most cases of application and further detailed analyses are not required. On the other hand, it is easy to notice that crystal–liquid phase equilibria and phase transitions (the terms phase transition and phase equilibrium are often employed here with a similar meaning except for the cases when the transformation kinetics is studied) are not as thoroughly analyzed as compared to the liquid–vapor phase transition. In the latter case, the existence of a critical point in the coexistence of two fluid phases defines characteristic scales of thermodynamic variables (volume, temperature, pressure, entropy and energy) and allows one to introduce the concepts of corresponding states and thermodynamic similarity of various substances.

For the crystal–liquid phase transitions of simple substances the situation is different. The melting lines were found not to contain with the increase of temperature a fundamental singularity like the liquid–vapor critical point. This feature makes impossible the natural choice of some scaling parameters similar to the liquid–vapor phase transition. But one can implement another approach to the problem, which is based on the low-temperature asymptotic behavior of the melting lines of substances of normal type. Such a procedure requires one to include into the thermodynamic consideration the behavior of the respective phases at metastable conditions. The mere fact of considering systems at such conditions represents one of the distiguishing features of the present book. Here the problem of similarities and differences of crystal–liquid and liquid–vapor phase transitions in single-component systems is the central problem under consideration. In this analysis, much attention is devoted to the revelation of the thermodynamic similarity in the behavior of different substances at the phase transitions. The reason for this is that similarity concepts exhibit the very general deep properties of a class of effects retaining some particular differences in behavior in other particular respects. The more complete is the understanding of the nature of the effects considered the more completely and clearly the similarity in the behavior can be exhibited. In the present work, the mentioned connection is clearly demonstrated in application both to liquid–vapor and crystal–liquid phase transitions first in application to the behavior of one-component systems. In the final chapter, the analysis is then extended to crystal–liquid–vapor and liquid–liquid phase equilibria in two-component systems analyzed from the same point of view.

The present work is written employing basically the framework of phenomenological thermodynamics. The application of statistical-mechanical approaches, e.g., by utilizing Gibbs's

Crystal–Liquid–Gas Phase Transitions and Thermodynamic Similarity. Vladimir P. Skripov and Mars Z. Faizullin

ISBN: 3-527-40576-3

canonical ensemble method, would have required the introduction of various assumptions and approximations into the analysis, inevitable in order to obtain concrete results. Here we concentrate on the analysis of the general features leaving detailed statistical-mechanical considerations to future investigations or referring to the existing literature.

1.2 States of Aggregation. Phase Diagrams and the Clausius–Clapeyron Equation

First-order phase transitions are characterized by a jump of the first-order derivatives of the Gibbs thermodynamic potential and by the existence of metastable states of each of the phases. An example is given in Fig. 1.1.

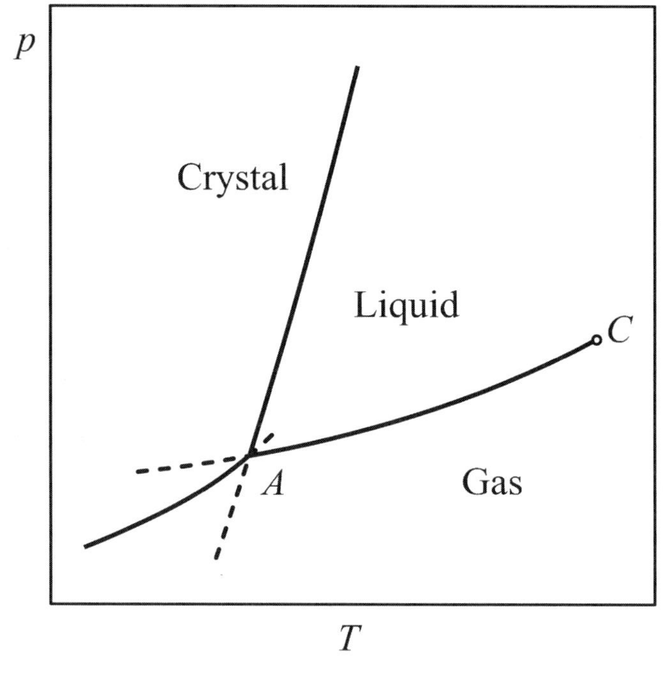

Figure 1.1: Phase diagram of a one-component system with crystal, liquid and vapor phases. By (C), the critical point of the liquid–vapor equilibrium is specified, (A) denotes the triple point. The dashed parts of the phase coexistence curves show their continuation into the respective metastable states.

Figure 1.1 shows a (p, T)-phase diagram of different states of aggregation of a single-component substance with a crystal phase, where the melting of the latter is characterized by a positive slope of the equilibrium crystal–liquid coexistence curve. The line of coexistence of liquid and vapor is terminated at the critical point denoted in the figure by C. In the critical point, the liquid and vapor phases become identical. The dashed parts of the curves show the extensions of phase equilibrium curves into the range of the respective metastable states of the coexisting phases.

The line of phase equilibrium for a homogeneous single-component system is defined by equality of the chemical potentials at the same values of temperature and pressure in both phases. For the liquid–vapor coexistence curve we have

$$\mu_L(T, p) = \mu_V(T, p) \, . \tag{1.1}$$

Here μ_L and μ_V are the chemical potentials of the liquid and the vapor, respectively, T is the temperature and p is the pressure.

By taking the derivative of Eq. (1.1) along the liquid–vapor equilibrium curve and, taking into account the relations $(\partial\mu/\partial T)_p = -s$, $(\partial\mu/\partial p)_T = v$, we get the Clausius–Clapeyron equation

$$\frac{dp}{dT_{LV}} = \frac{\Delta s_{LV}}{\Delta v_{LV}} \, , \tag{1.2}$$

where $\Delta s_{LV} = s_L - s_V$ and $\Delta v_{LV} = v_L - v_V$ are the jumps of specific entropy and volume in the phase transition. Both differences on the right-hand side of Eq. (1.2) are positive and the inequality $(dp/dT_{LV}) > 0$ holds.

Similarly one can write the Clausius–Clapeyron equation for crystal–liquid phase equilibrium. The entropy of the liquid is greater than the entropy of the crystal, $\Delta s_{SL} = s_L - s_S > 0$, so the slope of the melting line is determined by the sign of the difference Δv_{SL}. Substances, obeying the inequalities $\Delta v_{SL} > 0$ and $dp/dT_{SL} > 0$, are called normally melting. Here we will consider only such normally melting substances.

1.3 Metastable States. Relaxation via Nucleation

The curve as determined by Eq. (1.1) can be interpreted as a line of intersection of two surfaces in the (T, p, μ)-space. For both of these surfaces, this line is not a singular one. This property implies the possibility of a smooth extension of both phases into the regions of their metastable states. Figure 1.2 shows the trace of the surfaces for the crystalline $\mu_S(T, p)$ and liquid $\mu_L(T, p)$ phases on the plane $p =$ constant. The point O of the intersection of the lines L′L and SS′ corresponds to the phase equilibrium at the given pressure. For the parts OL′ and OS′ the chemical potential has a higher value than for the competing phase at the same values of T and p.

The relative stability of the phases is determined by the relation between the values of μ_S and μ_L. The more stable phase corresponds to the lower value of the chemical potential. The state of the phase, having at the given temperature and pressure higher values of the chemical potential, is called metastable. The phase in this condition is stable with respect to small (continuous) changes of the thermodynamic parameters and has a finite lifetime. Metastable states are unstable with respect to large-scale disturbances which lead to the formation of viable new phase nuclei. The metastable state is destroyed by nucleation and growth of the nuclei of a new phase which is more stable at the given values of temperature and pressure. In a system which is free of impurities initiating the phase transformation, nucleation takes place due to thermal fluctuations.

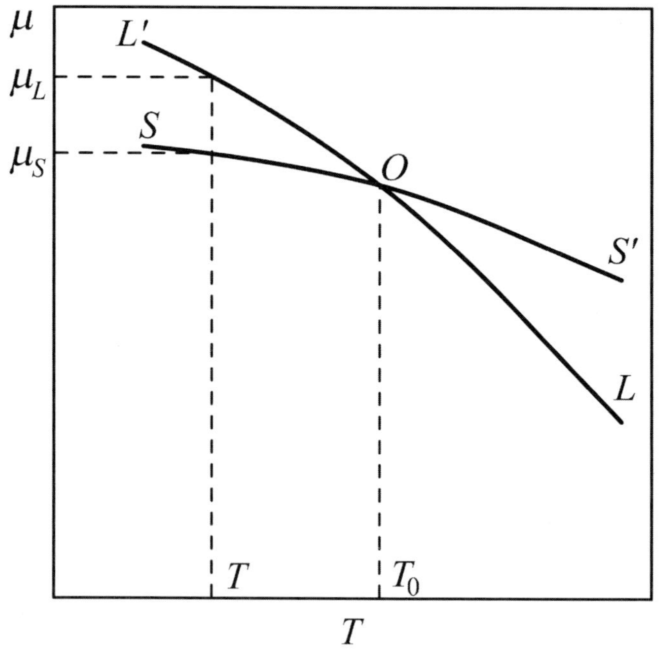

Figure 1.2: Behavior of the chemical potential of the crystalline (SS$'$) and liquid (LL$'$) phases in dependence on temperature on the surface p = constant near the temperature of the phase transition, T_0.

Metastable states and their quasi-static changes can be described by the equations of equilibrium thermodynamics. In order to allow such a description, the system should obey the following conditions [1]

$$\{t_i\} \ll t_{\text{exp}} < \bar{\tau} \,, \tag{1.3}$$

where t_i is the characteristic time of relaxation of the system under consideration with respect to the i-th state parameter (temperature, pressure, etc.), t_{exp} is the characteristic time of the experiment (the time required to transfer the system into the metastable state and to carry out the subsequent experimental observations), $\bar{\tau}$ is the mean waiting time for the formation of a nucleus of a more stable phase. The left part of the inequality in Eq. (1.3) guarantees the quasi-static character of the thermodynamic properties of the metastable phase. For such kinds of changes, the relations of equilibrium thermodynamics are fulfilled. This statement means that the system can be smoothly transformed into the metastable state without the occurrence of some kind of specific behavior in its properties at the point of the equilibrium phase transformation, if the system remains homogeneous. The latter condition is ensured by the right part of the inequality in Eq. (1.3).

The transfer of the system into metastable states is accompanied by a decrease of the thermodynamic stability of the respective phases. The equilibrium condition of the thermodynamic system (with respect to small (continuous) changes of the state parameters) requires the second variation of the specific internal energy $u(s,v)$ to be positive [2], i.e.

$$\delta^2 u = \left(\frac{\partial^2 u}{\partial s^2}\right)(\delta s)^2 + 2\left(\frac{\partial^2 u}{\partial s \partial v}\right)\delta s \delta v + \left(\frac{\partial^2 u}{\partial v^2}\right)(\delta v)^2 > 0 \,. \tag{1.4}$$

The inequality Eq. (1.4) holds true when the determinant, composed from the coefficients of the real-valued quadratic form Eq. (1.4), and its principal minors are positive

$$D = \begin{vmatrix} \dfrac{\partial^2 u}{\partial s^2} & \dfrac{\partial^2 u}{\partial s \partial v} \\ \dfrac{\partial^2 u}{\partial s \partial v} & \dfrac{\partial^2 u}{\partial v^2} \end{vmatrix} = \begin{vmatrix} \left(\frac{\partial T}{\partial s}\right)_v & \left(\frac{\partial T}{\partial v}\right)_s \\ \left(\frac{\partial T}{\partial v}\right)_s & -\left(\frac{\partial p}{\partial v}\right)_s \end{vmatrix} > 0 \,, \tag{1.5}$$

$$\left(\frac{\partial T}{\partial s}\right)_v = \frac{T}{c_v} > 0 \,, \qquad -\left(\frac{\partial p}{\partial v}\right)_s > 0 \,, \tag{1.6}$$

where c_v is the isochoric heat capacity. The derivatives in Eqs. (1.6) are called the adiabatic stability coefficients [3]. Zero values of the determinant D define the boundary of the thermodynamic phase stability with respect to continuous changes of the thermodynamic state parameters: this boundary is denoted as the spinodal.

The connection between the isodynamic partial derivatives and the stability determinant, D, is given by the following expressions [3]

$$\left(\frac{\partial T}{\partial s}\right)_p = -\frac{D}{(\partial p/\partial v)_s} \,, \tag{1.7}$$

$$-\left(\frac{\partial p}{\partial v}\right)_T = \frac{D}{(\partial T/\partial s)_v} \,. \tag{1.8}$$

The stability conditions Eqs. (1.5) and (1.6) are thus reduced to the positivity of the isodynamic partial derivatives

$$\left(\frac{\partial T}{\partial s}\right)_p = \frac{T}{c_p} > 0 \,, \tag{1.9}$$

$$-\left(\frac{\partial p}{\partial v}\right)_T = (v\beta_T)^{-1} > 0 \,, \tag{1.10}$$

which are also called isodynamic stability coefficients. Here c_p is the isobaric heat capacity and β_T is the isothermal compressibility. Zero values of the derivatives Eqs. (1.9) and (1.10) correspond to the spinodal of the system. Conditions (1.9) and (1.10) allow us to estimate the thermodynamic stability of the system and the distance to the spinodal by properties which may be obtained directly through experiment.

1.4 Phase Transformations in a Metastable Phase. Homogeneous Nucleation

The depth of entering the region of metastable states is naturally connected with the difference of the chemical potentials of the phases at given temperature and pressure. For example, for the supercooled liquid we have (see Fig. 1.2)

$$\Delta\mu_{SL} = \mu_L - \mu_S \; . \tag{1.11}$$

This value is a measure of the instability of the liquid phase and characterizes the driving force of the phase transformation. The difference of the chemical potentials, $\Delta\mu_{SL}$, can be expressed through the values of thermodynamic parameters which can be measured directly experimentally, i.e., via the corresponding deviation from the point of the equilibrium phase transition with respect to temperature, ΔT, and pressure, Δp [1]. For small deviations from phase equilibrium, the value $\Delta\mu_{SL}$ can be represented as

$$\begin{aligned}\Delta\mu_{SL} &= \left[\left(\frac{\partial\mu_L}{\partial T}\right)_p - \left(\frac{\partial\mu_S}{\partial T}\right)_p\right]\Delta T - \left[\left(\frac{\partial\mu_L}{\partial p}\right)_T - \left(\frac{\partial\mu_S}{\partial p}\right)_T\right]\Delta p \\ &= (s_L - s_S)\Delta T - (v_L - v_S)\Delta p \; .\end{aligned} \tag{1.12}$$

Here s and v are the specific entropy and volume at the temperature, T, and pressure, p, and the indices S and L refer to the crystal and the liquid, respectively. Expressions like Eq. (1.12) are also valid in cases when the depth of entering the metastable state region cannot be considered as small. In such cases, the values of entropy and volume for the crystal and the liquid are defined as mean values in the interval of supercooling.

The formation of a new phase in a metastable liquid is connected with overcoming the energetic barrier, the work of formation of a nucleus of critical size. If in a metastable system nuclei with sizes above the critical one are formed, then their further growth is governed thermodynamically. The process of formation of the first viable nucleus represents in this way the starting point of the phase transformation in the system. One way of describing processes of formation of clusters of supercritical sizes is homogeneous nucleation theory. Homogeneous nucleation theory presumes that the processes of formation and growth of new phase nuclei are the result of heterophase fluctuations proceeding by thermal fluctuations in the otherwise homogeneous ambient phase. The principal constituent parts of this theory were formulated first by Gibbs [4]. He was also the first to suggest that the thermodynamic stability of a metastable state should be related to the work of critical nucleus formation, W_*. Provided a critical vapor nucleus is formed in a metastable liquid, Gibbs's theory yields the following expression for the work of critical cluster formation

$$W_* = \frac{16\pi}{3}\frac{\sigma_{LV}^3}{(p'' - p')^2} \; , \tag{1.13}$$

where

$$p'' - p' = \frac{2\sigma_{LV}}{r_*} \tag{1.14}$$

is the difference of the pressure inside (p'') and outside (p') a spherical nucleus of critical size (with a radius r_*) and σ_{LV} is the surface tension of the liquid–vapor equilibrium interface. Eq. (1.14) is the condition of mechanical equilibrium in the superheated liquid–vapor system nucleus, denoted commonly as the Young–Laplace equation.

Volmer [5] has formulated the expression for the work of critical nucleus formation in the following widely equivalent way

$$W_* = \frac{16\pi}{3} \frac{\sigma_{LV}^3 v_V^2}{(\Delta\mu_{LV})^2} , \tag{1.15}$$

where, in application to boiling of liquids, v_V is the specific volume of the vapor phase, $\Delta\mu_{LV}$ is the difference of the chemical potentials of liquid and vapor phases at given pressure and temperature. In application to crystallization of liquids and taking into account Eq. (1.12) one can rewrite the expression for the work of critical crystal nucleus formation in a supercooled melt at given external pressure in the following form [6]

$$W_* = \frac{16\pi}{3} \frac{\sigma_{SL}^3 v_S^2}{(\Delta\mu_{SL})^2} = \frac{16\pi}{3} \frac{\sigma_{SL}^3 v_S^2}{(\Delta s_{SL})^2(\Delta T)^2} , \tag{1.16}$$

where σ_{SL} is the surface tension of the crystal–melt interface. In order to describe the state of an equilibrium crystallite in a melt, an effective surface tension is usually introduced by substituting the facetted shape of the crystal by a sphere of a radius r having the same volume as the crystal in its equilibrium shape [6]. The values v_S, Δs_{SL}, and σ_{SL} in Eq. (1.16) are supposed to be equal to the respective macroscopic state parameters of the newly evolving phase in equilibrium with the ambient phase. The above considerations are the essence of the thermodynamic approximation for the work of critical cluster formation in homogeneous nucleation theory.

The value W_* in Eq. (1.16) characterizes the height of the Gibbs free energy barrier which the system should overcome due to fluctuations. Increasing the depth of entering the region of metastable states, we decrease this barrier. Therefore, the probability of overcoming the barrier due to thermal fluctuations is increased. Dividing W_* by the mean energy of thermal motion per degree of freedom, k_BT, where k_B is the Boltzmann constant, one gets a natural dimensionless measure for the height of the barrier, denoted commonly as the Gibbs number, G_*

$$G_* = \frac{W_*}{k_BT} . \tag{1.17}$$

The Gibbs number, G_*, plays an important role in the theory of homogeneous nucleation. It determines to a large extent the value of the steady-state nucleation rate, J. The nucleation rate, J, is determined as the mean number of viable nuclei formed in a unit volume at unit time. For steady-state conditions, the expression for J can be represented in the following way

$$J = N_1 B_1 \exp(-G_*) , \tag{1.18}$$

where N_1 is the number of molecules in a unit volume of the liquid and B_1 is a kinetic pre-factor characterizing the mean rate of passing the range of near-critical nucleus sizes in cluster size space. For liquids with low viscosity, the parameter B_1 changes only slightly with an increase of the degree of supercooling, ΔT. By the order of magnitude, we have $N_1 \approx 10^{28}\,\mathrm{m}^{-3}$, $B_1 \approx 10^{10}$–$10^{11}\,\mathrm{s}^{-1}$. From Eqs. (1.16)–(1.18) we may conclude that a small change of supercooling, ΔT, of the liquid significantly influences the nucleation rate, J. Thus, the value of J is increased by 9 orders of magnitude for mercury (Hg) and by 4 orders of magnitude for tin (Sn) when the supercooling is increased by 10 K starting from $\Delta T = 52$ K (for Hg) and 122 K (for Sn) [6].

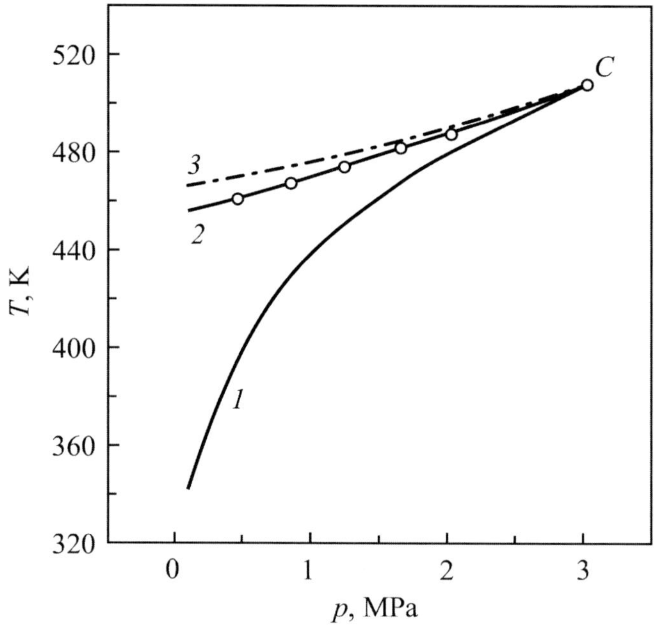

Figure 1.3: Binodal curve (1), line of accessible superheating (2) and spinodal curve (3) of hexane. The circles on curve (2) refer to experimental measurements of the accessible superheating, C denotes the critical point.

The searches for the detailed expressions for J were connected with the construction and the solution of the set of kinetic equations describing the evolution of an ensemble of nuclei of subcritical sizes. This way of formulating homogeneous nucleation theory is described, for example, in the works of Volmer [5], Zeldovich [7], and Frenkel [8]. The model, based on the diffusion model of the nucleation process, turned out to be universal, independent with respect to the nature of the phases capable of coexistence and metastability. This general character is proven, for example, by the systematic investigations of homogeneous nucleation in superheated liquids [1] and supercooled melts [6].

Equations (1.15)–(1.18) clearly exhibit the strong nonlinear dependence of the nucleation rate on the "driving force" of the phase transition, $\Delta\mu$. It results in a very fast increase of J with an increase of $\Delta\mu$. By this reason, the process of homogeneous nucleation exhibits features widely similar to some threshold phenomenon. Only after some certain degree of supersaturation is reached can perceptible nucleation be observed. Once this threshold is passed, the further increase of supersaturation leads to a dramatic increase of J.

Although the intensity of nucleation processes in a superheated liquid is increased with an approach to the spinodal, the degree of stability of the system with respect to discontinuous (heterophase) variations of its state is not directly determined by the stability of the system to relatively small homophase fluctuations. For processes of nucleation, connected with discontinuous (heterophase) fluctuations, the measure of stability is the mean waiting time, $\bar{\tau}$, for occurrence of a viable nucleus in a metastable system of a volume, V. This mean waiting time is connected with the nucleation rate via the relation [1,6]

$$\bar{\tau} = (J \cdot V)^{-1} . \tag{1.19}$$

It depends significantly on the interfacial tension, σ, and is not directly related to the isodynamic stability coefficients $-(\partial p/\partial v)_T$ and $(\partial T/\partial s)_p$.

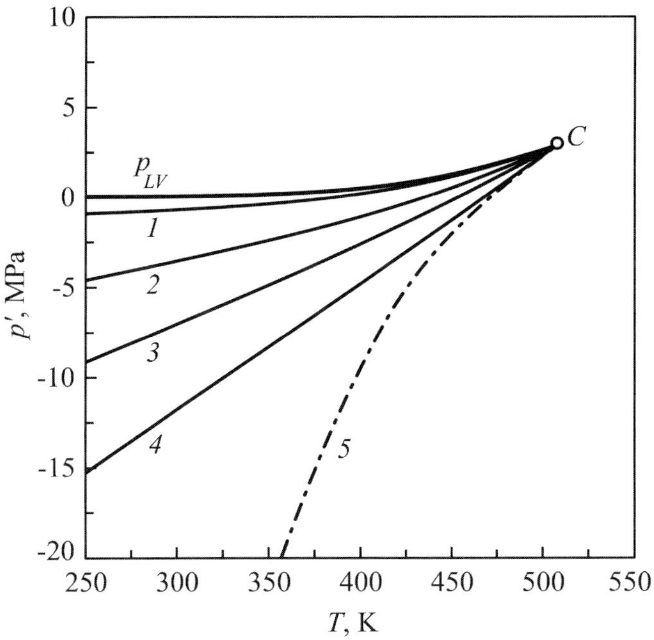

Figure 1.4: Family of curves corresponding to an unstable equilibrium for a heterogeneous state consisting of a cluster in the otherwise homogeneous ambient phase for different values of the radius, r, of a spherical bubble in superheated hexane. By p_{LV}, the macroscopic binodal is specified corresponding to $r \to \infty$. The other curves are drawn for the following values of r: (1) $r = 50$ nm, (2) $r = 10$ nm, (3) $r = 5$ nm, (4) $r = 3$ nm. Curve (5) is the spinodal curve of liquid hexane.

In Fig. 1.3, the binodal and the spinodal curves and the line of accessible superheatings for liquid hexane are shown corresponding to the steady-state nucleation rate $J = 10^6\ \mathrm{s}^{-1}\mathrm{cm}^{-3}$ [1]. Note that in supercooled single-component liquids there is no spinodal curve at all (see

Section 2.1); however, the nucleation kinetics is described by the same scheme as for the case of boiling of superheated liquids. The main condition of intensive spontaneous nucleation consists in a decrease of the nucleation barrier W_* to sufficiently low values, for example, when the Gibbs number is of the order

$$G_* = \frac{W_*}{k_B T} \approx 20\text{–}50 \; . \tag{1.20}$$

Except for the case of very thin layers, the surface tension does not enter the conditions of phase equilibrium of two phases coexisting at planar interfaces. In the order of magnitude, the ratio of the interfacial energy of the system and the bulk contributions depends on the size of the newly evolving phase, l, as $1/l$. For small particles ($l \approx r$) the surface energy gives a significant contribution to the conditions of their equilibrium with the environment (see Eq. (1.14)) and to the work of formation of an unstable equilibrium (critical) nucleus of the dispersed phase (cf. Eqs. (1.13) and (1.16)) in a sufficiently large volume of the parent (metastable) phase. If the radius of a new-phase particle, r, is taken as the size parameter, then one can plot a set of unstable equilibrium curves on a (T, p)-surface corresponding to different values of the size parameter, r. Each of these curves connects the temperature with the pressure of the external phase.

Figure 1.4 shows such a plot for gas bubbles in superheated hexane. With a decrease of r, the lines are shifted to the right along the temperature axis and to lower pressures as compared with the macroscopic binodal, p_{LV}. The latter curve corresponds to $(1/r) \to 0$. Note that for the line of accessible superheating of hexane (2), shown in Fig. 1.3, in the range $p > 0$ the radius of the critical nucleus increases with the increase of pressure (and temperature). By this reason, the line $J(T, p) =$ constant does not belong to the family of curves $r(T, p) =$ constant [1].

The phase boundary between the different phases consists, in general, of several molecular layers, its width, δ, is proportional to the correlation length. The minimal size of a new-phase particle also equals δ to an order of magnitude. The existence of a critical point in liquid–vapor equilibrium (or the liquid–liquid equilibrium in a solution) presupposes the growth of the correlation length and, consequently, also of the value of the parameter δ with the approach to the critical point. The absence of a critical point for crystal–liquid equilibrium in a single-component system makes the swelling of the transition layer impossible. Here its width is always limited in its extent to several intermolecular distances.

2 Liquid–Gas Phase Transitions

2.1 Basic Fact: Existence of a Critical Point

The ability of molecular systems, consisting of a large number of identical particles and being in thermal motion, to form various phases is a remarkable feature. The phases can coexist in equilibrium in a certain range of pressure and temperature. In the more general case, we have to consider multicomponent systems affected not only by external pressure, but also by other thermodynamic forces.

In a simple phase diagram, as shown in Fig. 1.1, phases coincide with the different states of aggregation, i.e., crystal, liquid and gas. In a single point (the triple point characterized by a temperature, T_A, and a pressure, p_A) these three states of aggregation are in equilibrium with each other at the corresponding values of the thermodynamic parameters of the coexisting phases $\rho = 1/v$ (density), s (entropy), and u (internal energy). In Fig. 2.1 the specific volumes v_S, v_L, and v_V in the triple point are specified with the letters A, A$'$, and A$''$. In addition, the spinodal curves for the crystal phase GH, the liquid CS$'$ and vapor CS$''$ are shown.

The critical point C is the point where the liquid–vapor phase equilibrium is terminated and the coexisting phases become identical. It is located both on the binodal and the spinodal curves as their common vertex (cf. Fig. 2.1). The bifurcation of the surface of the chemical potential, $\mu = \mu(T, p)$, similar to that shown (as a projection) in Fig. 1.2, is truncated with the increase of the pressure in the critical point (O→C). In the region above the critical temperature, the boundary of stability of the fluid state with respect to continuous perturbations ($D = 0$, $(\partial p/\partial v)_T = 0$) is not reached anywhere. The transition from the vapor to the liquid-like state at $T > T_c$ proceeds smoothly through a series of intermediate densities.

The discovery of a critical point in the liquid–vapor equilibrium curves as a universal feature of molecular systems was a remarkable fact in science of the 19th century. It established in theory the idea of the principal continuity of fluid states and, with respect to practical applications, it has supplied the governing idea underlying different methods of gas liquefaction. It has become clear in this way that strong compression alone is not sufficient to observe the coexistence of liquid and vapor phases, one has first to cool the gas below the thermodynamically determined critical temperature.

For a more detailed outline of the historical aspects of these problems, the interested reader is referred to the remarkable work "An overview on the development of our knowledge about gases" by A. G. Stoletov [9]. As outlined in this review, first a large group of substances was considered as non-condensable (permanent) gases. Step by step, the list of such substances shrank. For example, in 1823 the great physicist Michael Faraday could transfer into a liquid state nine substances considered earlier to be permanent gases, and 20 years later he could

Crystal–Liquid–Gas Phase Transitions and Thermodynamic Similarity. Vladimir P. Skripov and Mars Z. Faizullin

ISBN: 3-527-40576-3

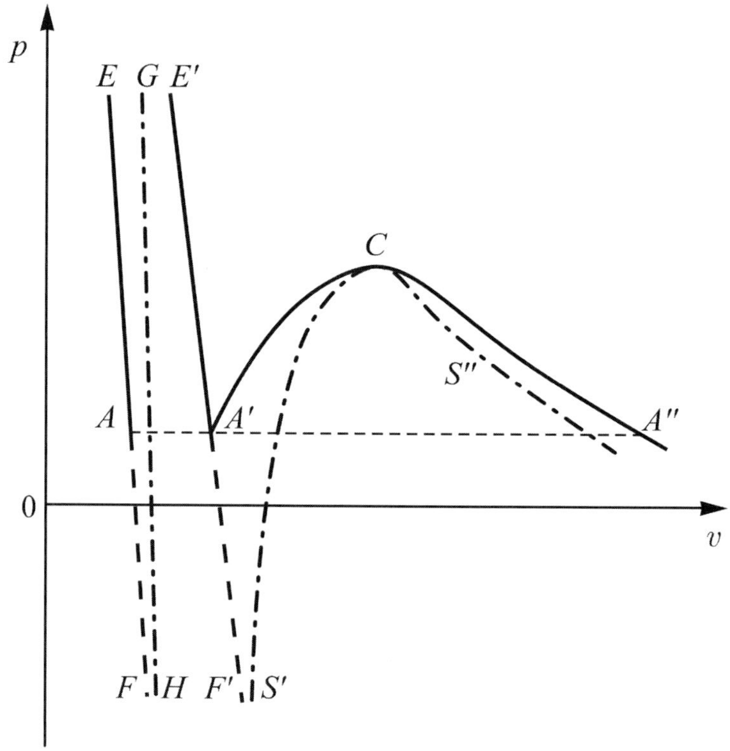

Figure 2.1: Phase diagram of a simple system: (A′CA″) is the binodal curve describing liquid–vapor equilibrium; (AE) and (A′E′) are the crystal and liquid branches of the crystal–liquid equilibrium curves; (S′CS″) is the spinodal of the liquid and the gas; (GH) is the spinodal of the crystal. The dashed parts of the curves correspond to the metastable continuation of the curves (EA) and (E′A′) beyond the triple point, (A–A′–A″). (C) is the critical point.

show the same for another six gases [9]. The events of such kind gave evidence of scientific progress, and they were reported to the interested public. For example, the following is the content of a telegram sent by Raul P. Pictet (1857–1937) from Geneva on January 11, 1878 (cited after [9]): "Yesterday, hydrogen was liquefied. 650 atmospheres and 140 degrees below zero. Hardened through evaporation. Intermittent stream of a steel-blue color, intensive fall-out of pellets onto the ground with an specific noise. Solid hydrogen remained in a tube for several minutes" (as it seems, in this note the initial values of pressure and temperature were given, ahead of liquefaction of hydrogen).

In the history of discovery of the critical point, the experiments of Cagnard de la Tour have played an important role (published in 1822, 1823, Ann. chim. et phys.). He heated liquids (ether, carbon disulfide, ethyl alcohol) in soldered glass tubes. Initially the liquid filled about one half of the tube, and the boundary with vapor was visible. Starting from a certain temperature, the boundary vanished and the substance became homogeneous. The normal separation into two phases was then restored with a decrease of temperature.

D. I. Mendeleev, while observing the decrease of surface tension, σ, of the liquid–vapor interface with an increase of temperature, concluded that there exists a certain temperature at which σ tends to zero, and "the liquid should turn into a body without cohesion, a gas, i.e., to become transformed into a vapor despite the small amount of space". Mendeleev called this temperature an absolute boiling point [10]. For ether and alcohol, Mendeleev's estimates of this temperature were close to the results of Cagnard de la Tour. Such a point of view served as the means to understand the critical temperature as a maximum limiting temperature for the existence of a liquid; however, such interpretation is wrong.

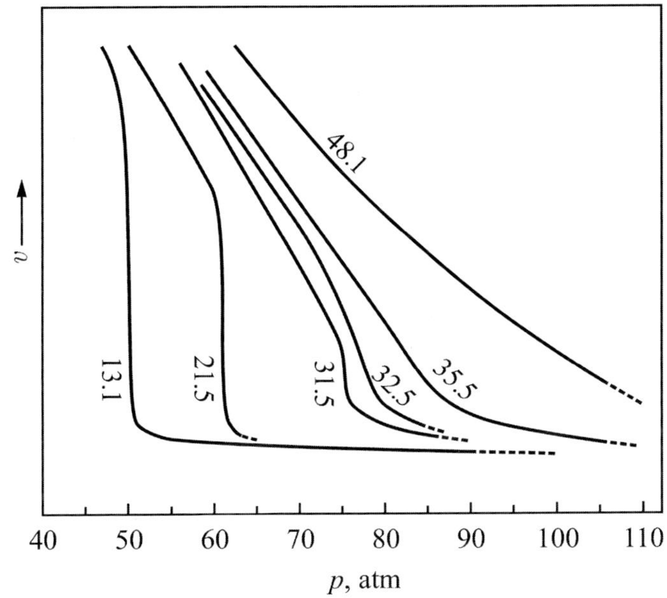

Figure 2.2: Isotherms of fluid states of carbon dioxide as obtained in Andrews's experiments.

The crucial contribution to the correct understanding of the essence of critical phenomena was established in the work of the Irish scientist Thomas Andrews (1813–1885). It was published in 1869 [11]. The author experimented with carbon dioxide at pressures up to 400 atmospheres. Carbon dioxide was compressed by mercury in a glass gauged tube 2.5 mm in diameter with a blind end. The tube was thermostated with water. Leather gaskets were used to tighten the joint with a brass head. The volume of air in a parallel-attached glass tube was used to measure the pressure. The pressure was created using a steel screw by screwing it into the water. Then, the pressure was transmitted via the mercury to the glass gauged tubes with CO_2 and air. For a series of temperatures (13.1; 21.5; 31.1; 32.5; 35.5; 48.1 °C), Andrews measured the change of the volume of CO_2 and air with the change of pressure. Employing this information, he plotted the $(p-v)$-isotherms shown in Fig. 2.2. In this figure, the ordinate axis gives the volume in relative units. In his subsequent article in 1876 [11], Andrews gave the values of the density, ρ, in g/cm^3. The main result of his analyses is that, on the basis of his experiments, Andrews formulated the statement about the continuity of the gaseous and liquid states of matter and that there has to exist a critical point above which an equilibrium of the two fluid phases cannot exist any more. For the critical temperature of CO_2, Andrews obtained the value 32.92 °C. This result is illustrated in Fig. 2.2. It shows that the vertical parts of the isotherms, which refer to two-phase conditions, shrink with the increase of temperature and tend to a point at the critical temperature, $T = T_c$. There exists a possibility of transition from the liquid to the gas without the loss of homogeneity of the system by-passing the critical point. For further analysis, it is interesting to quote from the first article of Andrews: "We have seen that the gaseous and liquid states are essentially only distinct stages of one and the same state of matter and that they are able to go over into each other by a continuous change. One much more complex problem remains thus to be solved – about the possible continuity of solid and liquid states of matter ..."

The work of Andrews gave powerful stimuli to further investigations. One of the first, who gave a comment, was D. I. Mendeleev. In 1870, he published his "Notes on the investigations of Andrews on the compressibility of carbon dioxide" where he pointed out that the critical temperature of Andrews is the same as the absolute boiling point introduced earlier by Mendeleev in 1860 [12].

An outstanding event in the development of science was the work of the Dutch physicist J. D. van der Waals (his dissertation published in 1873) entitled "The continuity of gaseous and liquid states" (this title coincides with the key phrase of Andrews's work; see the German translation [13]). In this thesis, the famous thermal equation of state

$$\left(p + \frac{a}{v^2}\right)(v - b) = RT \tag{2.1}$$

was proposed which is nowadays known as van der Waals's equation of state. Employing the conditions for the determination of the critical point

$$\left(\frac{\partial p}{\partial v}\right)_{T=T_c} = 0\,, \qquad \left(\frac{\partial^2 p}{\partial v^2}\right)_{T=T_c} = 0\,, \tag{2.2}$$

Eq. (2.1) can be transformed into a dimensionless form as

$$\left(\widetilde{p} + \frac{3}{\widetilde{v}^2}\right)(3\widetilde{v} - 1) = 8\widetilde{T}\,, \tag{2.3}$$

where the reduced variables $\widetilde{p} = (p/p_c)$, $\widetilde{T} = (T/T_c)$, and $\widetilde{v} = (v/v_c)$ are employed. The constants a, b, and R are connected with the critical parameters of a substance by the following expressions

$$v_c = 3b\,, \qquad p_c = \frac{1}{27}\frac{a}{b^2}\,, \qquad T_c = \frac{8}{27}\frac{a}{Rb} \tag{2.4}$$

or via

$$b = \frac{1}{3}v_c\,, \qquad a = 3p_c v_c^2\,, \qquad RT_c = \frac{8}{3}p_c v_c\,. \tag{2.5}$$

According to Eq. (2.3), the surface $\widetilde{p} = f(\widetilde{T}, \widetilde{v})$ has a shape as shown in Fig. 2.3. The critical isotherm passing through the point C divides the set of isotherms into two groups, the

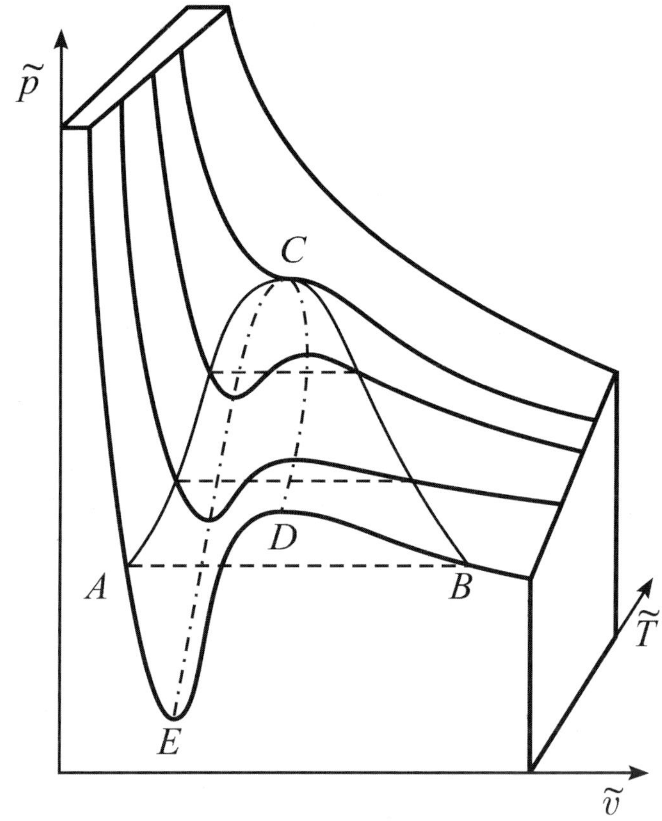

Figure 2.3: Thermodynamic surface describing the state of a one-component fluid according to van der Waals's equation of state, Eq. (2.3).

subcritical ones ($T < T_c$), each one having its minimum and maximum, and the supercritical ones ($T > T_c$) showing a monotonic behavior. Supercritical isotherms, which refer to temperatures near T_c, retain the memory about the vicinity of the phase separation region in the form of inflection points. The isotherms, plotted according to the van der Waals equation, retain continuity also below the critical point ($T < T_c, p < p_c$).

Without the application of some additional procedure, this equation does not lead to the liquid–vapor binodal (ACB in Fig. 2.3). This additional procedure, allowing one to determine the location of the binodal curve, was proposed by Maxwell. It is based on the equality of the chemical potential of both phases, $\mu_L(T, p) = \mu_V(T, p)$, at phase equilibrium. Maxwell's rule leads to the dashed horizontal lines (e.g., AB in Fig. 2.3), these lines corresponding to two-phase equilibrium states.

The spinodal curve ECD is composed of two branches, the liquid and vapor ones, going over into each other smoothly in the critical point. The parts of the isotherms located between the branches EC and DC of the spinodal curve correspond to unstable states; for such states, we have $(\partial p/\partial v)_T > 0$. Such states should be considered as thermodynamically unrealizable.

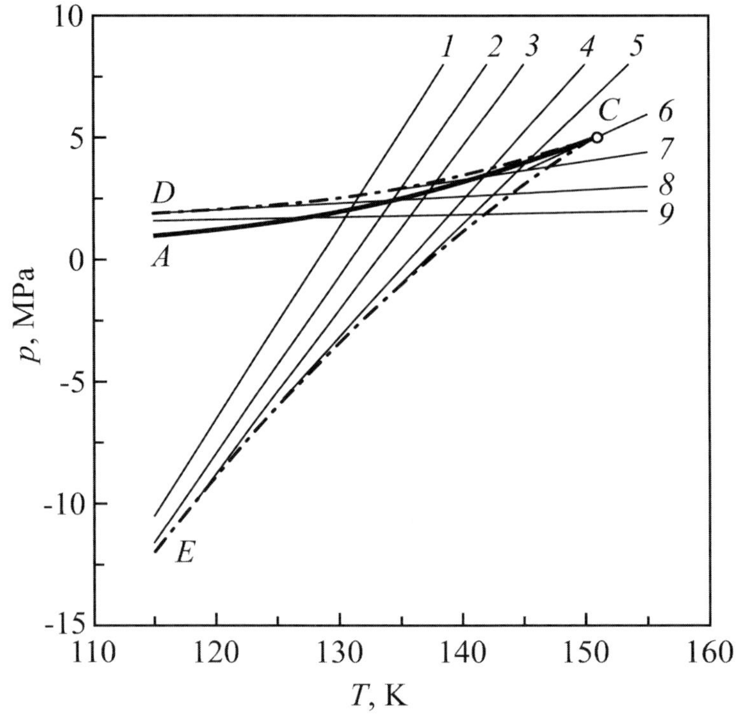

Figure 2.4: Phase diagram of the fluid states of argon: (AC) is the binodal, (EC) the spinodal of the liquid, (DC) the spinodal of the vapor, (1–5) is a family of isochores of the liquid, (6) is the critical isochore ($v_c = 1.867 \cdot 10^{-3}\ \mathrm{m}^3/\mathrm{kg}$), (7–9) are the isochores of the gas.

Figure 2.3 shows that binodal and spinodal curves are not planar ones, they are characterized by a complex location in the $(\widetilde{T}, \widetilde{v}, \widetilde{p})$-space. Some properties of the equation of state can be illustrated more accurately by using the projections of characteristic lines to one of the planes in (T, v, p)-variables. Let us consider, for example, the (T, p)-surface. If we plot according to Eq. (2.1) the spinodal branches CE and CD, the binodal AC and the set of isochores with their extension up to the spinodal, we can see an interesting fact: each of the branches of the spinodal is the envelope of the corresponding group of isochores. The plot is facilitated by the fact that the isochores $v(T, p) = \text{const.}$ are linear functions in van der Waals's equation since $(\partial p/\partial T)_v = R(v - b)$ holds. However, this property of the spinodal being an envelope of the corresponding family of isochores is generic [1] not being a specific feature of Eq. (2.1). In order to illustrate this statement, in Fig. 2.4 the respective plot for argon [14] is shown employing experimental (T, v, p)-data and extrapolations of the isochores beyond the binodal curve AC.

For the full thermodynamic description of the fluid, the thermal equation of state should be supplemented by a caloric equation of state containing the internal energy, u, or the heat capacity, c_v. For the fluid, determined by the van der Waals equation of state, Eq. (2.1), the

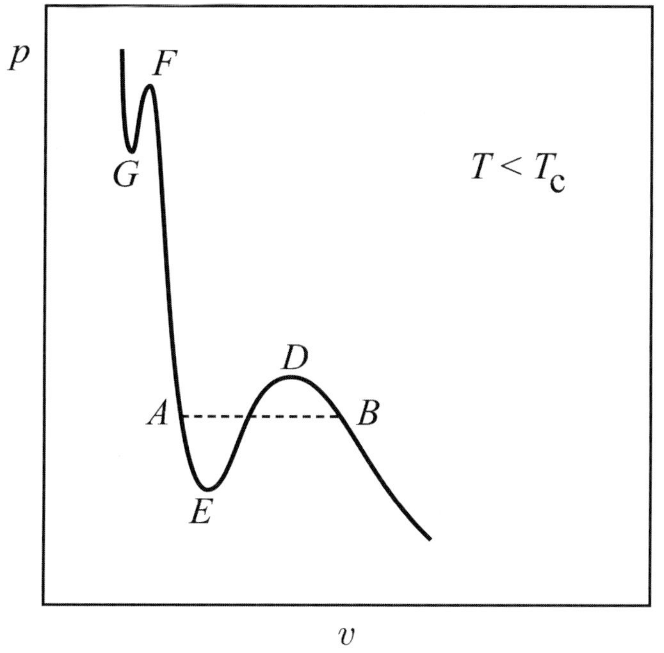

Figure 2.5: The course of the isotherm of the crystalline, the liquid and the vapor states of a one-component system if the existence of a spinodal of the undercooled (supercompressed) liquid is assumed. Note that such behavior is not found in reality (cf. Fig. 2.6).

heat capacity c_v does not depend on volume, since the relations

$$\left(\frac{\partial c_v}{\partial v}\right)_T = T\left(\frac{\partial^2 p}{\partial T^2}\right)_v = 0 \tag{2.6}$$

hold. This property facilitates the introduction of caloric parameters.

Formally the transition from the ideal gas equation, $pv = RT$, to Eq. (2.1) is performed by introducing two corrections. In one of them, to the external pressure the value of the internal pressure, (a/v^2), is added resulting from the intermolecular forces of attraction. The other correction takes into account the impermeability of the particles, i.e., the repulsion occurring at their contact. The type of attractive force, leading to the correction as introduced into the van der Waals equation, corresponds to a weak but long-range interaction of the molecules [15]. In such an approach, each particle is assumed to be located in a certain mean field of the forces resulting from its environment.

After the publication of van der Waals's work [13], a large number of different authors proposed their own equations of state [16] in order to reach a better agreement with experiment. However, significant advances in this respect were reached only by introducing a large number of parameters and resulted in a reduction of the clearness of the results. For the analysis of qualitative features of the liquid–vapor phase transition, the van der Waals equation of state remains the preferable one allowing one to demonstrate in a straightforward way the existence

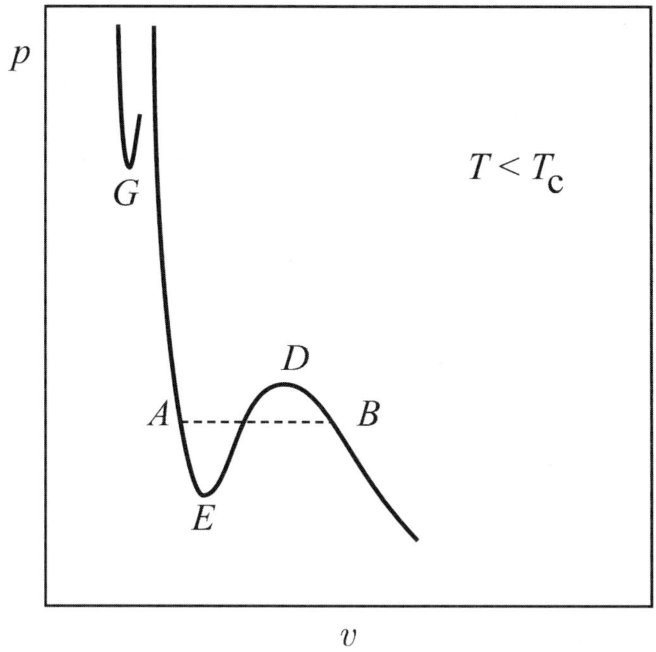

Figure 2.6: The course of the isotherms of the crystalline, the liquid and the vapor states of a one-component system for real systems as is observed in experiments. Such behavior is connected with the increase of the degree of stability of the undercooled (supercompressed) liquid with increasing degree of penetration of the system into the range of metastable states.

of the liquid–vapor equilibrium critical point and the existence of metastable homogeneous states located in between binodal and spinodal curves.

The success of the van der Waals equation of state in explaining, at least, the essential qualitative features of liquid–vapor phase equilibria may lead to the impression that a proper account of both the intermolecular forces of attraction and repulsion may be sufficient in order to describe qualitatively not only vapor–liquid phase transitions, but also liquid–crystal phase equilibria. Following such an idea, in some books the subcritical isotherms ($T < T_c$) are extended to densities corresponding to the crystal, and a second van der Waals-type loop is drawn as is shown in Fig. 2.5 [17].

However, the commonly employed continuous equations do not lead to such behavior of the isotherms. If we follow the course of the isotherms as shown in Fig. 2.5, then we should accept the existence not only of a spinodal point E in the expanded liquid, but also of the spinodal point F in a supercompressed liquid. In contrast, experiments prove that the elasticity coefficient $-(\partial p/\partial v)_T$ of the liquids beyond the condensation line AB steadily increases along the isotherms with an increase of pressure. As a result, the degree of stability of the liquid increases, and there is no evidence of approaching some spinodal point F. This conclusion is confirmed, for example, also by results of computer modeling of argon [18].

Going the opposite direction, i.e., by expanding the system in the crystalline state, its stability is decreased and a spinodal point G can really be reached.

According to the above analysis, the course of the curves as shown in Fig. 2.5 has to be changed in a way as shown in Fig. 2.6. The difference of these figures is of principal importance and is in agreement with the concept of the absence of a crystal–liquid equilibrium critical point and the impossibility of a smooth transition between these two states of aggregation of matter. If this conclusion – as we believe – is true, then there cannot exist a universal continuous equation of state describing the different states of aggregation (i.e., of vapor, liquid, crystal) of simple substances in a correct way. The crystalline state has to be described by a separate (T, v, p)-equation, and the melting line is found from the condition of equality of the chemical potential of the system in the solid (S) and liquid (L) states, i.e.,

$$\mu_S(T,p) = \mu_L(T,p) \ . \tag{2.7}$$

2.2 Method of Thermodynamic Similarity

The van der Waals equation in dimensionless variables Eq. (2.3) does not contain individual constants and is a particular expression of the so-called law of corresponding states for various substances. States with the same values of the variables $\widetilde{T}$, $\widetilde{v}$, and $\widetilde{p}$ are called the corresponding states, and the substances described by Eq. (2.3) are called thermodynamically similar. According to above given definitions, the general form of the law of corresponding states for the description of thermal properties of fluid systems is reduced to the identity of the thermal equations of states (in reduced variables) like

$$\varphi(\widetilde{p}, \widetilde{T}, \widetilde{v}) = 0 \tag{2.8}$$

for various substances [19]. It can be shown that the law of corresponding states holds true for substances with properties which can be described by an equation of state containing no more than two individual constants [20].

As an example of thermodynamic similarity, in Fig. 2.7 the behavior of the reduced density is shown along the saturation line, $\widetilde{\rho}_{LV} = \rho_{LV}/\rho_c$, of liquid and vapor for a group of simple substances. It follows from the figure that the points representing different substances can be described with a good approximation by the same dependence $\widetilde{\rho}_{LV}(\widetilde{T})$. The thermodynamic similarity of different substances in the region of liquid–vapor states is a reflection of the similarity of the intermolecular forces. By this reason, the law of corresponding states in the form of Eq. (2.3) is valid only for a small group of substances with van der Waals-type interactions.

A more general description of the similarity requires the introduction of additional parameters into Eq. (2.8). These parameters have to reflect the influence of individual peculiarities of the substances on the relationship between the variables $\widetilde{p}$, $\widetilde{T}$, and $\widetilde{v}$. As it is shown experimentally, a one-parameter equation of state of the form

$$\varphi(\widetilde{p}, \widetilde{T}, \widetilde{v}, A) = 0 \tag{2.9}$$

satisfactorily reflects the properties of gases and liquids for a wide class of nonassociated substances [21]. The parameter A in Eq. (2.9) has the meaning of a measure of similarity.

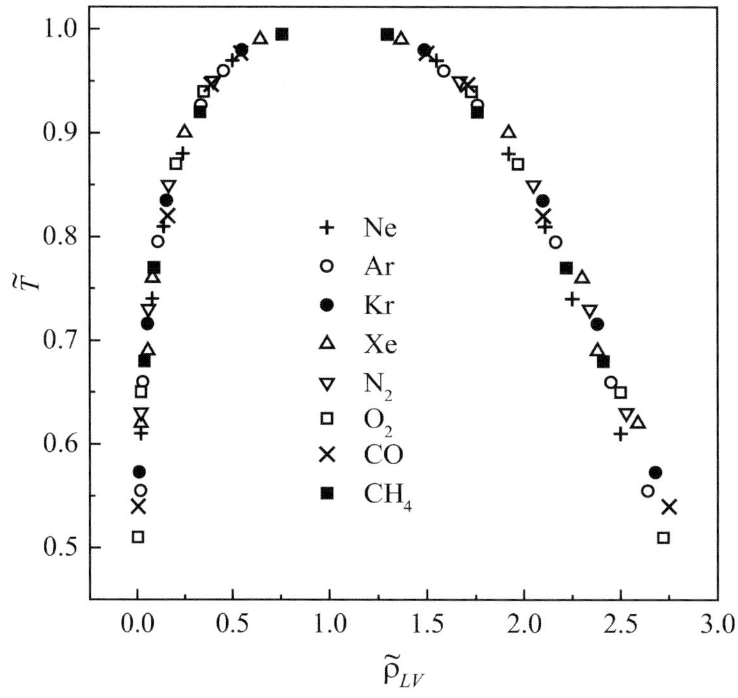

Figure 2.7: Dependence of the reduced density of liquid and vapor along the saturation curve on reduced temperature for simple substances.

Substances with equal or nearly equal values of A form a group of thermodynamically similar substances.

The parameter A can be introduced in different ways. For instance, in Ref. [21] the similarity parameter is introduced via the reduced pressure on the saturation line $\widetilde{p}_{LV}$ as

$$A = 100\widetilde{p}_{LV} \qquad \text{for} \qquad \widetilde{T} = 0.625 \, . \tag{2.10}$$

Employing the representation of Eq. (2.10), the parameter A has the highest value ($A = 4$) for inert gases forming a group of thermodynamically similar substances. For substances for which the shape of the molecules is significantly different from spherical, the values of the parameter A are usually low, e.g., for eicosane, we have $A = 0.2$.

In the work of Pitzer [22], the following definition of a similarity parameter ω_P was proposed

$$\omega_P = -\lg \widetilde{p}_{LV} - 1 \qquad \text{at} \qquad \widetilde{T} = 0.7 \, , \tag{2.11}$$

which was denoted by him as eccentricity criterion. Another similarity parameter, the Riedel similarity parameter ω_R, is defined by the slope of the saturation line in the critical point of the substance [23]

$$\omega_R = \frac{d \ln \widetilde{p}_{LV}}{d \ln \widetilde{T}} \, . \tag{2.12}$$

Using an appropriate approximation for the dependence of saturated vapor pressure on temperature, $p_{LV} = p_{LV}(T)$, one can easily obtain inambiguous relationships between the parameters A, ω_P and ω_R as [21]

$$\omega_R = 4.919\omega_P + 5.811 , \qquad \omega_P = 0.401 - 0.664 \lg A . \tag{2.13}$$

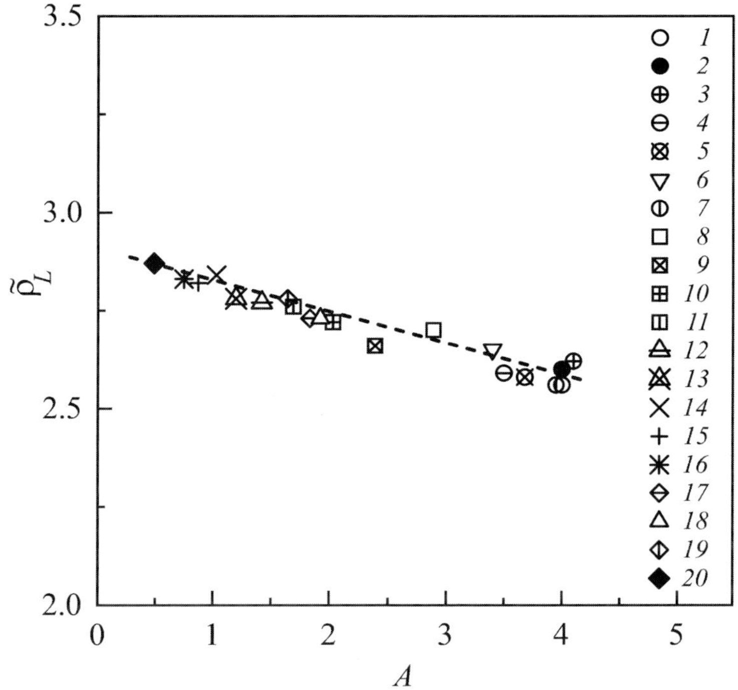

Figure 2.8: Dependence of the reduced density of the liquid on the parameter A along the saturation curve at a reduced temperature, $\widetilde{T} = 0.6$, for different substances: (1) argon, (2) krypton, (3) xenon, (4) nitrogen, (5) oxygen, (6) carbon monoxide, (7) methane, (8) ethane, (9) propane, (10) butane, (11) pentane, (12) hexane, (13) heptane, (14) octane, (15) nonane, (16) decane, (17) cyclohexane, (18) benzene, (19) toluene, (20) perfluorooctane.

In contrast to Eq. (2.8), expressions like Eq. (2.9) combine different groups of thermodynamically similar substances having equal (or, at least, close) values of the parameter A. The discovery of universal relationships between various different thermodynamic properties for the groups of substances, based on the analysis of experimental data, is an important task in the application of the method of thermodynamic similarity.

The curves showing the reduced densities of liquid and vapor along the saturation line ($\widetilde{\rho}_L(\widetilde{T})$ and $\widetilde{\rho}_V(\widetilde{T})$) form a family of nonintersecting curves which can be ordered employing the similarity criterion, A, as defined above. For the density of the liquid at the temperature $\widetilde{T} = 0.6$, the dependence $\widetilde{\rho}_L(A)$ is shown for a group of substances in Fig. 2.8. The analysis of experimental data, taking into acount the discovered dependence $\widetilde{\rho}_L(A)$, allows one to determine the form of the one-parameter relationship $\widetilde{\rho}_L(\widetilde{T}, A)$. For the substances represented in Fig. 2.8, this relationship has the following form [21]

$$\widetilde{\rho}_L = 1 + 0.85(1 - \widetilde{T}) + (2.09 - 0.654 \lg A)(1 - \widetilde{T})^{1/3} . \tag{2.14}$$

A similar relationship can be obtained also for the dependence of the density of the saturated vapor on temperature for this group of substances.

The one-parameter approximation Eq. (2.9) holds true for a wide range of states of liquid and vapor ranging from ideal gases to dense liquids close to the melting line. It has been shown [14] that there are no physical limitations to also employing it in the region of metastable states.

2.3 Similarity Near the Critical Point: The Change of Critical Indices

The applicability of thermodynamic similarity concepts in the vicinity of the liquid–gas critical point is of particular interest. As a rule, in the analysis of the behavior of different systems in the vicinity of T_c one is interested in the dependence of the thermodynamic properties of the system under consideration on the dimensionless parameter, $\tau = (T - T_c)/T_c$, or some other quantity reflecting the vicinity of the state of the system to the critical temperature. The change of the respective properties written in a dimensionless form is then represented as some exponential function of τ. For example, the isothermal compressibility at the critical isochore

$$K_T = -\frac{1}{v}\left(\frac{\partial v}{\partial p}\right)_T = \frac{1}{\rho}\left(\frac{\partial \rho}{\partial p}\right)_T \tag{2.15}$$

can be written as $p_c K_c = \Gamma_0 |\tau|^{-\gamma}$. Similarly, the dependence of the pressure on density on the critical isotherm is determined by the exponent δ

$$(p - p_c)/p_c = D_0 \left|\frac{(\rho - \rho_c)}{\rho_c}\right|^{\delta} , \tag{2.16}$$

$$(\tilde{p} - 1) = D_0 |\tilde{\rho} - 1|^{\delta} . \tag{2.17}$$

The difference of the densities of coexisting liquid and vapor is governed by a parameter β

$$\tilde{\rho}_L - \tilde{\rho}_V = B_0(-\tau)^{\beta} . \tag{2.18}$$

In the approach to the critical point C, simultaneously with the increase of the intensity of density fluctuations, the correlation length, ξ, is increased [24]. The asymptotic behavior of the correlation length, ξ, at $\rho = \rho_c$ and $\tau > 0$ is determined by the exponent ν via $\xi \sim \tau^{-\nu}$. Similarly to the relations given above, several other properties may be represented [24, 25]. The self-consistency of the different relations and the connections between different critical indices are verified from relations being consequences from thermodynamic stability conditions.

The critical indices may be determined by experiment. The ensemble of their values corresponds to one or the other model of behavior of the system under consideration in the vicinity of the critical point. In the classical (van der Waals) mean-field approach, we arrive at $\beta = 0.5$, $\gamma = 1$, $\nu = 0.5$, etc. The van der Waals equation Eq. (2.1) serves as one of the interpolation equations in the mean-field model [2, 15]. Experimentally it was observed, however, that with the approach to the critical point the critical indices change and have different values as compared with the classical ones, for example, $\beta = 0.34$, $\gamma = 1.25$, $\nu = 0.63$, etc. A systematic analysis of such crossover from a classical to a non-classical behavior of the ensemble of critical indices was performed in the 1960s and 1970s. At the same time, also the basic new ideas in the theoretical analysis of critical phenomena were developed. The history and the state of the art of the problem can be found in various monographs [15, 24, 25]. In any of the possible attempts to treat the respective phenomena, the increase of the intensity of fluctuations and the increase of the correlation length, ξ, near the critical point have to be accounted for. In the nonclassical approaches (denoted as Ising or renormalization group approaches) it is commonly assumed as an essential ingredient of the theory that – in the vicinity of the critical point – the particular form of the short-range interaction potentials does not affect the long-range order correlations. In the determination of the behavior of the system, some cooperative effects due to the interactions of fluctuations of different length scale dominate.

A detailed analysis of critical phenomena is out of the scope of the present monograph; this is a separate large topic. Here we concentrate on one particularly interesting point, i.e., on the analysis of the problem of what the effect of an external force field is on critical phenomena in the vicinity of the critical point [26]. Indeed, experiments directed to the determination of the critical indices are commonly performed in the gravitational field of the earth where the acceleration, g, is the measure of the field strength. Due to the increase of the compressibility in the vicinity of the critical point, in any measuring cell of finite height, z, the system becomes heterogeneous with respect to the density distribution. The lower layers of the substance in the cell are subjected to some excess pressure due to the weight of the upper layers of the substance. Each of such layers of width, dz, results in an additional contribution, dp, given by the relation $dp = -\rho g\, dz$. The excess density is small at small values of z; however, the change of the density of the substance,

$$d\rho = -\rho g \left(\frac{\partial \rho}{\partial p}\right)_T dz \,, \tag{2.19}$$

may be of essential significance. For the local density gradient, we obtain

$$|\nabla \vec{\rho}| = \left|\frac{d\rho}{dz}\right| = \rho^2 K_T g \,. \tag{2.20}$$

In the determination of the critical indices by experiment, such changes of the density of the substance have to be taken into account. This problem became especially important in connection with first analyses [27–29] reporting on a return after the first crossover from classical to nonclassical behavior to classical values of the set of critical indices in the immediate vicinity of the critical point in the course of decreasing the parameter τ. The question arises whether such second crossover is eventually the result of the above characterized effect of change of the density due to the high compressibility of the fluids near the critical point. A comprehensive analysis of the critical behavior in the presence of external fields is given in the Ref. [30].

In Ref. [29], a thorough experiment for the determination of the compressibility of two pure substances (SF_6 and CO_2) was performed employing two measuring cells of different

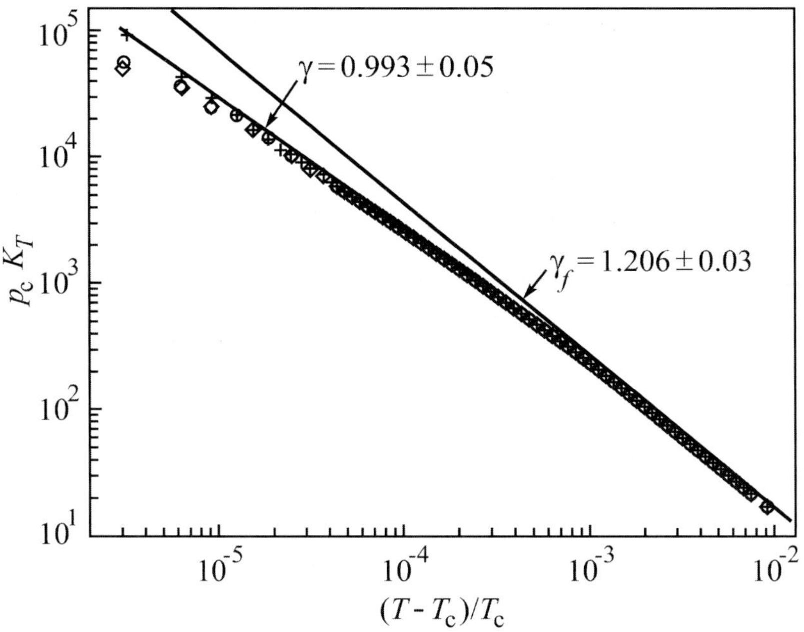

Figure 2.9: Dependence of the dimensionless isothermal compressibility on the dimensionless temperature along the critical isochore for SF_6.

sizes with a height $z = 30$ mm and $z = 11$ mm, respectively. The authors came to the conclusion [29] about the reality of the second crossover, i.e., the transition from Ising-like to classical behavior with respect to the critical exponent γ along the critical isotherm. The respective results are illustrated in Fig. 2.9 based on the data given in Ref. [29]. In Fig. 2.9, the dimensionless isothermal compressibility, $p_c K_T$, is shown in dependence on the dimensionless temperature. The position where the second crossover occurs, is indicated as well. A detailed consideration of the critical behavior of systems in the presence of exciting fields is given in Ref. [30].

One can suppose with sufficient foundation that the spatial inhomogeneity of the system near the critical point is not only an accompanying effect of the influence of external fields but an essential feature having a significant impact on the results and determining the return to classical critical indices in the immediate vicinity of the critical point. Taking into account this effect, it is possible to understand the mechanism of the second crossover assuming that a highly correlated fluctuational state of the substance (corresponding to a nonclassical Ising-type behavior) is incompatible with the existence of externally induced strong density gradients. In this way, in order to realize a non-classical behavior, the two conditions

(i) small deviations from the critical point (small values of τ) and

(ii) small density gradients

have to be fulfilled. For the establishment of the classical (mean-field) type of critical behavior, both conditions are of no relevance.

In systems of other nature as compared with liquid–vapor coexistence, like ferromagnetics and solutions, the role of the density is played by other thermodynamic parameters (magnetization, concentration or chemical potential). The fulfillment of both conditions (i) and (ii) facilitates for sufficiently low values of τ the first crossover (transition from classical to Ising-type behavior). In the presence of perturbing fields, with the further decrease of τ the second condition cannot be fulfilled any more and the system returns to the classical type of behavior (second crossover). We will demonstrate below how such behavior is connected with the reaction of the system to the external field at a given strength of the field. As it turns out, the system leaves, at the decrease of τ, the region of states with Ising-type behavior in a self-organized trajectory.

For these purposes, let us return to the expression for the density gradient Eq.(2.20) along the critical isochore. As some scaling parameters, we introduce the fluctuational density gradient

$$(\nabla \rho)_{Sc} = \frac{\rho_c}{\xi} \tag{2.21}$$

and the number

$$St = \frac{|\nabla \rho|}{(\nabla \rho)_{Sc}} , \tag{2.22}$$

which characterizes the density stratification. At $\rho = \rho_c$, we have

$$St = \rho_c K_T \xi g . \tag{2.23}$$

In agreement with condition (ii), the number (St) has to fulfill the inequality

$$St < a . \tag{2.24}$$

The number a is not defined so far; however, from Eqs. (2.23) and (2.24) we may observe that its value depends on the strength of the gravitational field, g. Strictly speaking, the value of the parameter a is not a constant but depends also on τ. Assuming the values of the critical indices to be equal to $\gamma = 1.25$ and $\nu = 0.63$, we get the following expression for the temperature dependence of the characteristic number, St

$$St = Ag\rho_c\tau^{-1.9} . \tag{2.25}$$

It is evident that for sufficiently small values of τ the conditions Eq. (2.24) cannot be fulfilled any more, in other words, the trajectory leaves the Ising-range of critical behavior as illustrated in Fig. 2.10. In this figure, the full curve shows the boundary $a = a(\tau, g = \text{const.})$. Note that $|\nabla \rho|$ and $(\nabla \rho)_{Sc}$ approach each other with a decrease of τ: The term $|\nabla \rho|$ increases according to $\tau^{-\gamma}$, and the term $(\nabla \rho)_{Sc}$ decreases as τ^{ν}. If in an experiment the condition $St > a$ is fulfilled for all possible values of τ, the system always shows a classical behavior and avoids entering areas with Ising-type behavior. Such behavior is found, for example, at very large values of the field strength, g. Having in mind these considerations, we can reformulate the conditions (i) and (ii) for nonclassical behavior in the following form: intensive highly correlated fluctuations have to develop at small field-induced gradients.

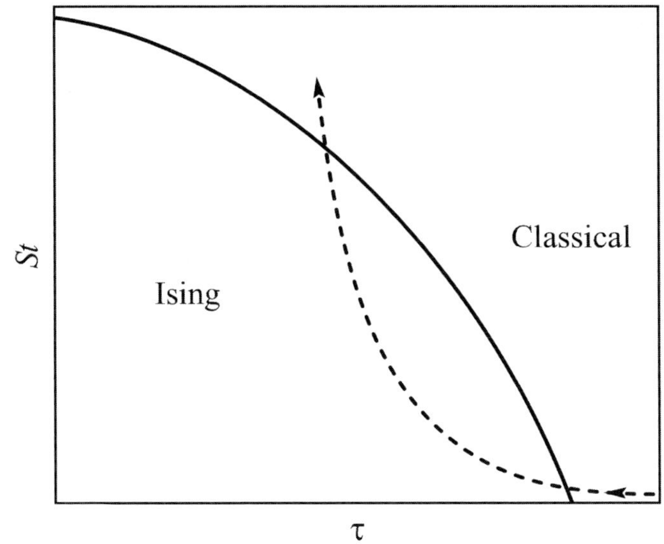

Figure 2.10: Boundary between the areas of classical and nonclassical behavior of a system under the influence of an external field.

Completing the analysis, we would like to draw attention to an interesting analogy. The (T, H)-phase diagram of a superconductor of first kind is similar to the behavior illustrated in Fig. 2.10. The curves, dividing the areas of normal and superconducting behavior, may be approximately expressed via the dependence [31]

$$\frac{H}{H_0} = 1 - \left(\frac{T}{T_0}\right)^2 . \tag{2.26}$$

At $T = T_0$, the magnetic field destroying superconductivity tends to zero and for $T \to 0$ the critical value of the magnetic field has its maximum equal to H_0. A normal conductor may exist practically at any value of the magnetic field H; however, in order to find the substance in a superconducting state, the condition

$$H < H_c = H_0 \left[1 - \left(\frac{T}{T_0}\right)^2\right] \tag{2.27}$$

has to be fulfilled. Taking into account such properties, one may bypass the area of superconducting behavior maintaining a magnetic field of sufficiently high field strength, $H > H_c(T)$. Varying the field strength H along the trajectory, one may also enter first the area of superconducting behavior at $T = T_0$, and then leave it again by decreasing the temperature. It follows that one may realize a trajectory similar to that shown in Fig. 2.10. In a superconductor, there is no process of self-organization of the trajectory of the field H to its value $H_c(T)$ at $T < T_0$. In a fluid near the critical point, such self-organization takes place; at $T \to T_c$ we find an increase of the gradient of the scalar field $\rho(z)$ (resulting in an increase of the parameter St)

due to the increase of the susceptibility of the system to the external impact. At $H \neq 0$, the transition into a superconducting state proceeds via a phase transition of first order. The difference between a van der Waals-fluid- and an Ising-fluid-type behavior is a weak one: they correspond to two different branches in the approach of the critical state.

2.4 New Universal Relationships for Liquid–Vapor Phase Coexistence in One-Component Systems

The application of the method of thermodynamic similarity has turned out to be fruitful also for a description of thermophysical properties of substances in the region of liquid–vapor states. In the framework of a one-parameter approach many relationships have been discovered allowing us to calculate with a good precision the properties of gases and liquids in a wide range of states including metastable ones [14,21,32]. Such relationships are found to be useful for predictive calculations, including the selection of new substances and materials by the given properties [33].

When the new generalized variables, characterizing the phase equilibrium, are introduced the analysis of two-phase fluid systems allows us to discover not previously observed universal relationships between dimensionless thermodynamic parameters which do not require the introducion of a similarity parameter A. This feature enlarges significantly the picture of thermodynamic similarity in application to liquid–vapor phase transitions.

2.4.1 Correlation Between Pressure and Densities of Liquid and Vapor Along the Saturation Curve

In order to understand at a qualitative level the behavior of a two-phase liquid–vapor system let us consider, again, the van der Waals equation of state, Eq. (2.1). For a fluid described appropriately by such an equation of state, the expressions for the molar internal energy, u, and the molar entropy, s, have the following form [34]

$$u = c_v T - \frac{a}{v} , \qquad s = c_v \ln T + R \ln(v - b) . \tag{2.28}$$

The heat capacity at constant volume, c_v, of the van der Waals fluid does not depend on volume. Employing this result and Eqs. (2.28), one can obtain a relationship for the chemical potential, μ, which can be expressed generally as

$$\mu = u - Ts + pv . \tag{2.29}$$

Then, taking into account the condition of phase equilibrium for liquid and vapor Eq. (1.1), we get the following well-known expression for the equilibrium pressure

$$p_{LV} = \frac{RT}{v_V - v_L} \ln\left(\frac{v_V - b}{v_L - b}\right) - \frac{a}{v_L v_V} . \tag{2.30}$$

With the notation

$$p_*(T) = \frac{a}{v_L v_V} = a\rho_L\rho_V \ , \tag{2.31}$$

and employing the expansion of $\ln(1 - b/v)$ with respect to the parameter $x = b/v$ (considering it as small), one obtains the following equation

$$\frac{p_{LV}}{p_*(T)} + 1 \cong \frac{RT}{a}\left[\frac{\ln\left(\dfrac{\rho_L}{\rho_V}\right)}{\rho_L - \rho_V} + b\right] . \tag{2.32}$$

Using the definition Eq. (2.31), the value $p_*(T)$ in Eq. (2.32) can be interpreted as a certain weighted mean internal pressure at the interval (ρ_L, ρ_V). It varies from zero at $T = 0$ (on the metastable extension of the saturation line beyond the triple point) to (a/v_c^2) in the critical point. The structure of Eq. (2.32) shows that, in the van der Waals model, the equilibrium pressure of the vapor is defined mainly by the value $p_*(T)$ via Eq. (2.31), i.e., by the product of orthobaric densities of liquid and vapor.

Employing experimental sets of data, we find that for a group of different substances the product $T[\ln(\rho_L/\rho_V)/(\rho_L - \rho_V)]$ decreases from the triple point to the states near the critical point approximately by 20%, while the product $\rho_L\rho_V$ is increased by four orders of magnitude. The expression of Eq. (2.32), obtained strictly for a van der Waals fluid only, can be considered thus as a certain "hint" allowing us to expect the existence of a universal dependence of the form $p_{LV} = f(\rho_L\rho_V)$ for a wide class of substances.

The analysis of experimental data concerning the pressure of saturated vapor, p_{LV}, and the orthobaric densities, ρ_L and ρ_V, for a wide class of substances has shown [35,36] that the power law dependence

$$p_{LV} = F\left(\rho_L\rho_V\right)^n \tag{2.33}$$

with $n = 1.1$ approximates quite well the experimental data in the range of temperatures from the triple point to $T/T_c \approx 0.9$. The factor F in Eq. (2.33) is an individual constant for the specific substance analyzed. With respect to pressure, the upper boundary of the interval of validity of this equation corresponds to a pressure $p/p_c \approx 0.5$. These results are illustrated in Fig. 2.11. This figure shows on a logarithmic scale the dependence between saturation pressure and the product $\rho_L\rho_V$ for water and mercury for a series of temperatures increasing in the positive direction of the ordinate. A deviation from linearity is observed only with an approach to the liquid–vapor critical point.

In order to establish the universal character of Eq. (2.33), we have to turn to the reduced quantities $\widetilde{p}_{LV} = p_{LV}/p_c$, $\widetilde{\rho}_L = \rho_L/\rho_c$, and $\widetilde{\rho}_V = \rho_V/\rho_c$. In reduced variables, Eq. (2.33) can be rewritten as

$$\widetilde{p}_{LV} = \widetilde{F}\left(\widetilde{\rho}_L\widetilde{\rho}_V\right)^n \qquad \text{with} \qquad \widetilde{F} = F p_c^{-1}\rho_c^{2n} \ . \tag{2.34}$$

Figure 2.12 shows experimental data in such reduced variables for substances of different nature, while Table 2.1 contains the values of n and $\widetilde{F}$ for these substances calculated by

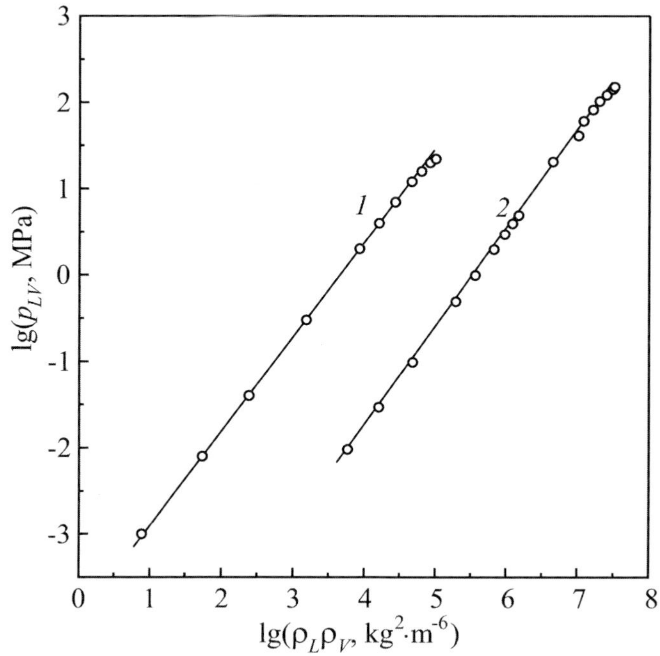

Figure 2.11: Dependence of the saturation pressure, p_{LV}, on the product of the densities of the liquid, ρ_L, and the vapor, ρ_V, for water (1) and mercury (2).

using least-square fitting methods. Figure 2.12 and Table 2.1 prove the universal character of the dependence between the product of orthobaric densities and the saturation pressure in a wide range of temperatures. Employing Eq. (2.34), we can set $n = 1.13 \pm 0.02$ and $\widetilde{F} = 1.3 \pm 0.1$.

With an approach to the critical point the character of the dependence changes but remains universal. For the interval $0.9 < \widetilde{T}_{LV} \leq 1$, where $\widetilde{T}_{LV} = T/T_c$, experimental data are described by the following expression:

$$\widetilde{p}_{LV} = \ln\left(\widetilde{\rho}_L\widetilde{\rho}_V\right)^m + 1\ . \qquad (2.35)$$

The mean value of the exponent m for different substances is 0.65. Figure 2.13 illustrates the dependence, given by Eq. (2.35), for a group of substances.

Equations (2.34) and (2.35) reflect the universal connection between the values p_{LV} and $(\rho_L\rho_V)$. The average difference between theoretical predictions and experiment for various substances is less than 1%. Tables 2.2 (for water) and 2.3 (for xenon) show the respective experimental data of the saturation pressure, $\widetilde{p}_{LV(exp)}$, and of the values $\widetilde{p}_{LV(calc)}$ calculated by using Eqs. (2.34) and (2.35) based on the data for the orthobaric densities of liquid and vapor. A comparison of the values $\widetilde{p}_{LV(exp)}$ and $\widetilde{p}_{LV(calc)}$ proves that these equations describe the dependence $\widetilde{p}_{LV} = f(\widetilde{\rho}_L\widetilde{\rho}_V)$ with a good precision.

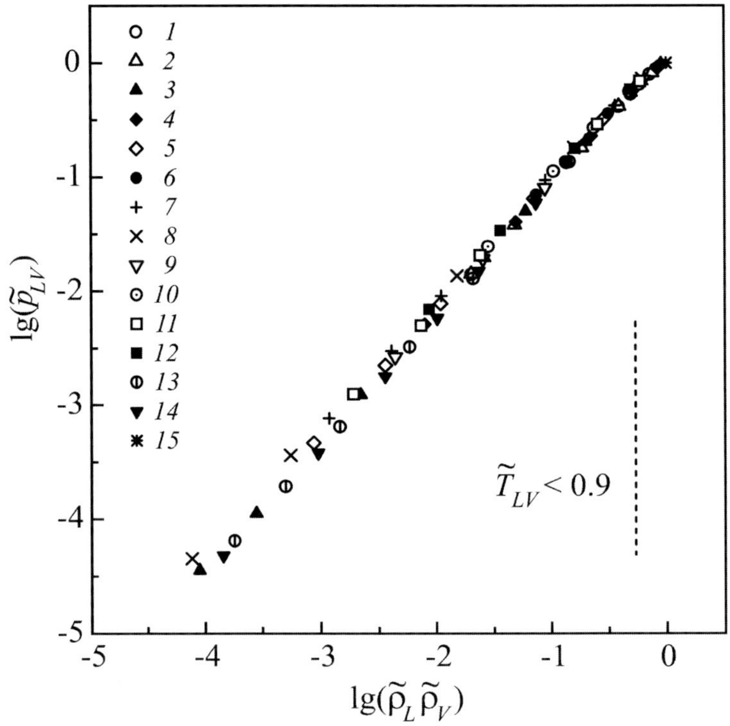

Figure 2.12: Dependence of the reduced pressure, $\widetilde{p}_{LV}$, on the product, $(\widetilde{\rho}_L \widetilde{\rho}_V)$, for substances of different nature: (1) argon, (2) xenon, (3) oxygen, (4) nitrogen, (5) chlorine, (6) carbon dioxide, (7) ammonia, (8) water, (9) methane, (10) hexane, (11) benzene, (12) diethyl ether, (13) mercury, (14) cesium; (15) specifies the location of the critical point.

2.4.2 Correlation Between Caloric Properties and Densities of Liquid and Vapor Along the Saturation Curve

Using the van der Waals equation, one can represent as a function of the product $\rho_L \rho_V$ the behavior of the caloric properties characterizing the liquid–vapor phase transformation [46], i.e., the entropy of the phase transition $\Delta s_{LV} = s_V - s_L$, the enthalpy difference $\Delta h_{LV} = h_V - h_L = T\Delta s_{LV}$ and the internal energy change $\Delta u_{LV} = \Delta h_{LV} - p_{LV}\Delta v_{LV}$. Indeed, if we introduce the entropy (cf. Eq. (2.28)) of the different phases into the condition of phase equilibrium (Eq. (1.1)) and take into account that the heat capacity does not depend on volume, we obtain

$$p_{LV} = T\frac{\Delta s_{LV}}{\Delta v_{LV}} - a\rho_L \rho_V \tag{2.36}$$

or

$$\frac{\Delta h_{LV}}{\Delta v_{LV}} - p_{LV} = a\rho_L \rho_V \ , \qquad \frac{\Delta u_{LV}}{\Delta v_{LV}} = a\rho_L \rho_V \ . \tag{2.37}$$

Table 2.1: Parameters of Eqs. (2.34) and (2.40) for various substances. The references to the literature, from where the experimental data for p_{LV}, ρ_L, ρ_V, and Δh_{LV} are taken, are given in the last column.

Substance	n	$\widetilde{F}$	k	H^*	Ref.
Argon	1.15	1.25	0.97	0.98	[37]
Xenon	1.14	1.23	0.97	1.00	[37]
Oxygen	1.13	1.23	0.99	0.98	[38]
Nitrogen	1.13	1.23	0.97	0.99	[39]
Chlorine	1.13	1.30	0.98	0.98	[40]
Carbon dioxide	1.12	1.31	0.95	1.00	[40]
Ammonia	1.12	1.38	0.98	0.97	[40]
Water	1.10	1.33	0.97	0.98	[41]
Methane	1.14	1.25	0.98	1.01	[42]
Hexane	1.11	1.34	0.97	1.01	[40]
Benzene	1.14	1.40	0.97	0.99	[40]
Diethyl ether	1.11	1.35	0.96	1.01	[40]
Mercury	1.14	1.13	0.99	–	[40, 43]
Cesium	1.14	1.09	1.00	0.99	[44, 45]

Therefore, for a van der Waals fluid, the value of the combination of thermodynamic parameters

$$\varphi = \frac{\Delta h_{LV}}{\Delta v_{LV}} - p_{LV} \tag{2.38}$$

is proportional to the product $\rho_L \rho_V$.

Assuming that in the general case the dependence analyzed is of the form

$$\varphi = H\left(\rho_L \rho_V\right)^k , \tag{2.39}$$

we calculate the values of φ and $\rho_L \rho_V$ for a set of substances using experimental data in a wide temperature range along the phase equilibrium curve. Figure 2.14 shows the dependence of φ on $\rho_L \rho_V$ on a logarithmic scale for water, argon, cesium, and mercury. In each of these cases, the points are located on a straight line with a coefficient k with $k \approx 1$ provided the inequality $T/T_c \leq 0.9$ holds.

It is convenient to transfer in Eqs. (2.38) and (2.39) to dimensionless variables taking some reference state along the saturation line, for example, a point with a pressure $p_1 = 0.1 p_c$. In terms of the variables $\varphi^* = \varphi/\varphi_1$ and $\rho_L^* \rho_V^* = (\rho_L \rho_V)/(\rho_L \rho_V)_1$, Eq. (2.39) takes the following form

$$\varphi^* = H^* (\rho_L^* \rho_V^*)^k , \tag{2.40}$$

where $H^* = H \varphi_1^{-1} (\rho_L \rho_V)_1^k$ holds.

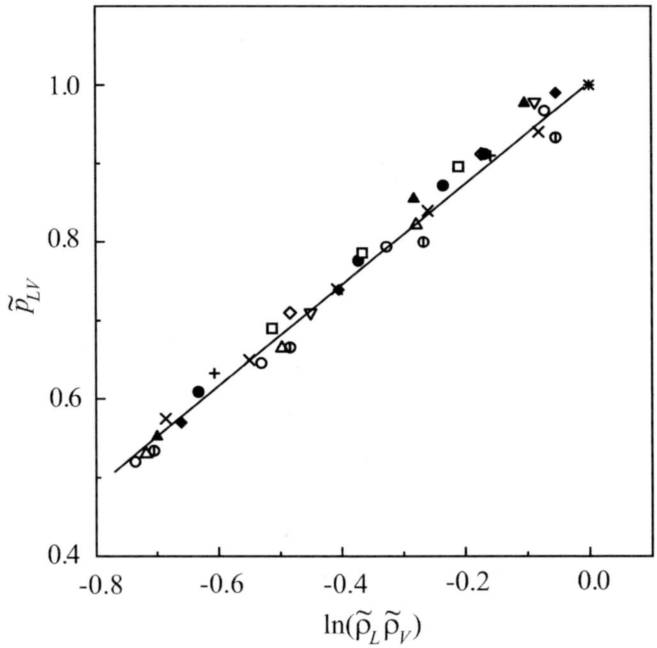

Figure 2.13: Dependence of the reduced pressure, $\widetilde{p}_{LV}$, along the coexistence curve on the product, $(\widetilde{\rho}_L \widetilde{\rho}_V)$, in the range of reduced temperatures, $0.9 \leq \widetilde{T}_{LV} \leq 1$, for the same substances (and with the same symbols) as given in Fig. 2.12. The curve corresponds to Eq. (2.35).

Figure 2.15 shows the experimental data for $\rho_L^* \rho_V^*$ and φ^* for the same substances as presented in Fig. 2.12 except mercury. It is evident that we get a confirmation of the universal character of the multiplicative expression Eq. (2.40) for the ratios $\Delta h_{LV}/\Delta v_{LV}$ and $\Delta u_{LV}/\Delta v_{LV}$, containing caloric parameters. The value of the exponent k is $k \approx 1.0$ and the amplitude H^* is also $H^* \approx 1.0$. The more precise values of the parameters k and H^* for various substances are given in Table 2.1.

In the vicinity of the critical point, in the temperature interval $0.9 < \widetilde{T}_{LV} \leq 1$, the dependence φ^* on $\rho_L^* \rho_V^*$ can be described by the following expression

$$\varphi^* = \ln \left(\rho_L^* \rho_V^*\right)^r + E \tag{2.41}$$

with mean values of the parameters of $r = 4.3$ and $E = -2.5$. The behavior of this dependence for various substances is shown in Fig. 2.16. It follows from the figure that the universal character of the behavior of $\varphi^*(\rho_L^* \rho_V^*)$ is retained in the temperature range $0.9 < \widetilde{T}_{LV} \leq 1$. The scattering of the points at $T \to T_c$ can be explained by the low accuracy in the determination of the ratio $\Delta h_{LV}/\Delta v_{LV}$ in Eq. (2.38) for φ in the immediate vicinity of the critical temperature.

The function φ (cf. Eq. (2.38)) can be represented in an alternative equivalent form as

$$\varphi = T \frac{\Delta s_{LV}}{\Delta v_{LV}} - p_{LV} = \frac{\Delta u_{LV}}{\Delta v_{LV}} . \tag{2.42}$$

Table 2.2: Comparison of experimental pressures $\widetilde{p}_{LV}$ [41] with the theoretical estimates calculated by using Eqs. (2.34) and (2.35). The data given here are for water.

T, K	p_{LV}, MPa	$\widetilde{p}_{LV(exp)}$	$\widetilde{p}_{LV(calc)}$	$\dfrac{\widetilde{p}_{LV(exp)} - \widetilde{p}_{LV(calc)}}{\widetilde{p}_{LV(exp)}}$	Equation
280.13	0.001	0.0000452	0.0000436	0.035	Eq. (2.34)
306.05	0.005	0.000226	0.000228	−0.007	
314.68	0.008	0.000361	0.000366	−0.015	
349.04	0.040	0.00181	0.00186	−0.029	
406.69	0.300	0.0136	0.0138	−0.015	
485.52	2.00	0.0904	0.0893	0.012	
506.99	3.00	0.136	0.133	0.019	
523.48	4.00	0.181	0.178	0.018	
558.95	7.00	0.316	0.314	0.006	
576.49	9.00	0.407	0.409	−0.006	
584.11	10.0	0.452	0.457	−0.011	
597.79	12.0	0.543	0.547	−0.007	Eq. (2.35)
609.78	14.0	0.633	0.639	−0.009	
620.47	16.0	0.723	0.725	−0.003	
630.11	18.0	0.814	0.807	0.009	
638.86	20.0	0.904	0.891	0.014	
642.94	21.0	0.954	0.934	0.021	
644.91	21.5	0.972	0.958	0.015	
647.27	22.115	1	1	0	

Another expression for φ can be obtained by using the Clausius–Clapeyron equation. In this case, we can write

$$\varphi = T\left(\frac{dp}{dT}\right)_{LV} - p_{LV} \,. \tag{2.43}$$

In Tables 2.4 (water) and 2.5 (xenon) experimental data of the quantities $\varphi^*_{(exp)}$ are compared with the values $\varphi^*_{(calc)}$ calculated by using Eqs. (2.40) and (2.41) based on the data for the orthobaric densities of the liquid and the vapor along the boundary curve. Comparing $\varphi^*_{(exp)}$ and $\varphi^*_{(calc)}$, one can see the highly universal character of the obtained correlations. The difference between the calculations and experiment for various substances is less than about 1%. The relationships thus established can be used to estimate the evaporation enthalpy based on the knowledge of the orthobaric densities at different temperatures.

One can suppose that the nature of the established universal character of the dependence of the heat of evaporation on the product of orthobaric densities of liquid and vapor on the equilibrium curve, similarly as for the dependence of saturation pressure p_{LV} on $\rho_L\rho_V$ (see Eqs. (2.34) and (2.35)), results from the fundamental properties of the interacting particles (molecules) in the presence of forces of both attraction and repulsion between them. This

Table 2.3: Comparison of experimental pressures $\widetilde{p}_{LV}$ [37] with the theoretical estimates calculated by using Eqs. (2.34) and (2.35). The comparison is performed here for xenon.

T, K	p_{LV}, MPa	$\widetilde{p}_{LV(exp)}$	$\widetilde{p}_{LV(calc)}$	$\frac{\widetilde{p}_{LV(exp)} - \widetilde{p}_{LV(calc)}}{\widetilde{p}_{LV(exp)}}$	Equation
161.36	0.0816	0.0140	0.0141	−0.007	Eq. (2.34)
180	0.2218	0.0381	0.0383	−0.005	
190	0.3480	0.0598	0.0600	−0.003	
200	0.5212	0.0895	0.0893	0.002	
210	0.7504	0.129	0.128	0.006	
220	1.045	0.179	0.178	0.007	
230	1.416	0.243	0.241	0.008	
240	1.872	0.321	0.320	0.003	
250	2.425	0.417	0.418	−0.002	
260	3.087	0.530	0.526	0.008	Eq. (2.35)
265	3.462	0.595	0.600	−0.008	
270	3.869	0.665	0.672	−0.011	
275	4.310	0.740	0.743	−0.004	
280	4.786	0.822	0.816	0.007	
285	5.299	0.910	0.888	0.024	
289.74	5.821	1	1	0	

Table 2.4: Comparison of the experimental data for φ^* [41] and the theoretical values calculated by using Eqs. (2.40) and (2.41). The data in this table are for water.

T, K	p, MPa	$\varphi^*_{(exp)}$	$\varphi^*_{(calc)}$	$\frac{\varphi^*_{(exp)} - \varphi^*_{(calc)}}{\varphi^*_{(exp)}}$	Equation
280.13	0.001	0.000971	0.00102	−0.049	Eq. (2.40)
314.68	0.008	0.00710	0.00678	0.046	
349.04	0.040	0.0288	0.0288	0.0	
406.69	0.300	0.175	0.171	0.023	
485.52	2.00	0.918	0.899	0.020	
523.48	4.00	1.671	1.660	0.007	
558.95	7.00	2.710	2.748	−0.014	
584.11	10.0	3.707	3.839	−0.035	
597.79	12.0	4.366	4.431	−0.015	Eq. (2.41)
609.78	14.0	5.005	5.131	−0.025	
620.47	16.0	5.660	5.779	−0.021	
630.11	18.0	6.360	6.392	−0.005	
638.86	20.0	7.060	7.027	0.005	
642.94	21.0	7.450	7.353	0.013	
644.91	21.5	7.680	7.540	0.018	

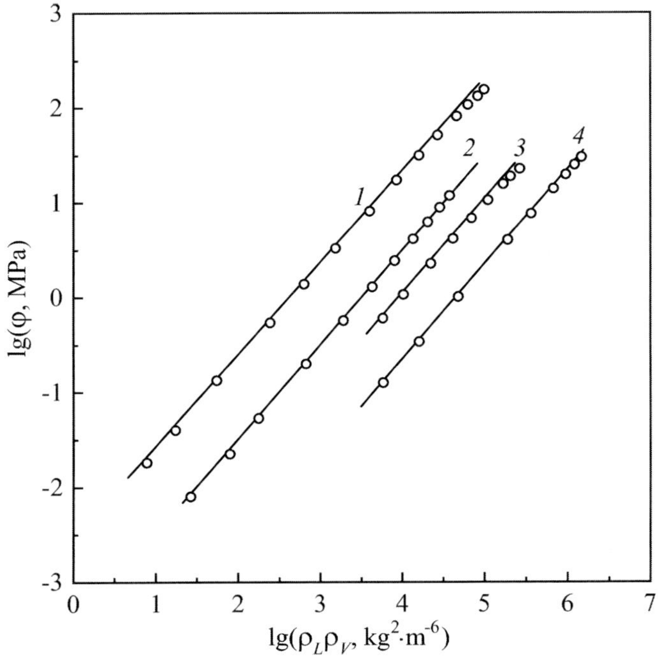

Figure 2.14: Dependence of the parameter $\varphi = (\Delta h_{LV}/\Delta v_{LV}) - p_{LV}$ on the product of the densities of the liquid and the vapor along the boundary curve: (1) mercury, (2) water, (3) cesium, (4) argon.

conclusion is supported by the fact that a simple account of the forces resulting in the correction terms in the van der Waals equation, allows one to establish an explicit relation between p_{LV}, Δh_{LV} and the product $\rho_L \rho_V$ defining the pressure p_* (cf. Eq. (2.31)) at phase equilibrium.

2.4.3 Correlation Between Surface Tension and Heat of Evaporation of Nonassociated Liquids

Molecular-kinetic concepts applied to the understanding of the surface properties of liquids indicate the existence of a relation between the surface tension, σ_{LV}, and the specific heat of evaporation, Δh_{LV}. An analysis of experimental data made it possible to establish a number of empirical relationships between these properties [47,48]. The use of the thermodynamic similarity methodology for describing the surface tension of normal (nonassociated) liquids resulted in the discovery of a one-parameter relationship of the form [49,50]

$$\sigma^*_{LV} = \sigma^*_{LV}\left(\Delta h^*_{LV}, A\right) \tag{2.44}$$

with

$$\sigma^*_{LV} = \sigma_{LV} k_B^{-1/3} T_c^{-1/3} p_c^{-2/3}\,, \qquad \Delta h^*_{LV} = \frac{\Delta h_{LV}}{RT}\,. \tag{2.45}$$

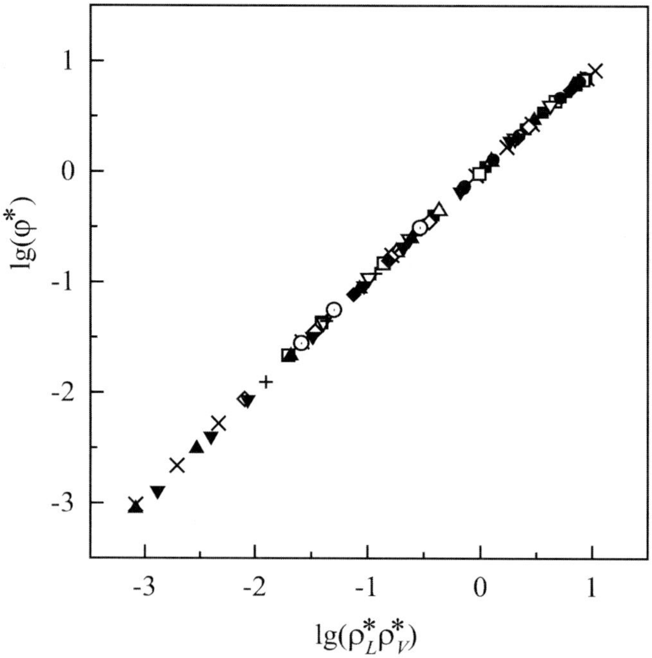

Figure 2.15: Dependence of the reduced quantity φ^* along the saturation curve on the product $\rho^*_L \rho^*_V$ for the same group of substances as presented in Fig. 2.12.

Here k_B is the Boltzmann constant and A is the thermodynamic similarity parameter, which characterizes the individual features of the molecules similarly to Eq. (2.10). The analysis of the relation between the value of σ_{LV} and the heat of evaporation per unit volume of the liquid phase, $\Delta h_{LV}/v_L$, allows us to establish new universal correlations not containing individual parameters for the description of normal liquids [51].

Let us introduce dimensionless variables for σ_{LV}, Δh_{LV}, and v_L, taking as scaling parameters the corresponding values at the temperature $\widetilde{T}_{LV} = T/T_c = 0.6$

$$\widetilde{\sigma}_{LV} = \frac{\sigma_{LV}}{(\sigma_{LV})_{0.6}} , \qquad \Delta\widetilde{h}_{LV} = \frac{\Delta h_{LV}}{(\Delta h_{LV})_{0.6}} , \qquad \widetilde{v}_L = \frac{v_L}{(v_L)_{0.6}} . \tag{2.46}$$

Figure 2.17 shows the dependence between the quantities $\widetilde{\sigma}_{LV}$ and $\Delta\widetilde{h}_{LV}/\widetilde{v}_L$ for a wide class of nonassociated liquids. Table 2.6 shows the scaling values for the calculation of the dimensionless variables defined via Eq. (2.46). A separate column in Table 2.6 contains the references to the literature, from where the experimental data for Δh_{LV} and v_L and the critical parameters T_c and p_c were taken. As follows from Fig. 2.17, the points for various substances match well the general curve in a wide range of values of the variables. The origin of the system of coordinates corresponds to the critical point of the respective substances.

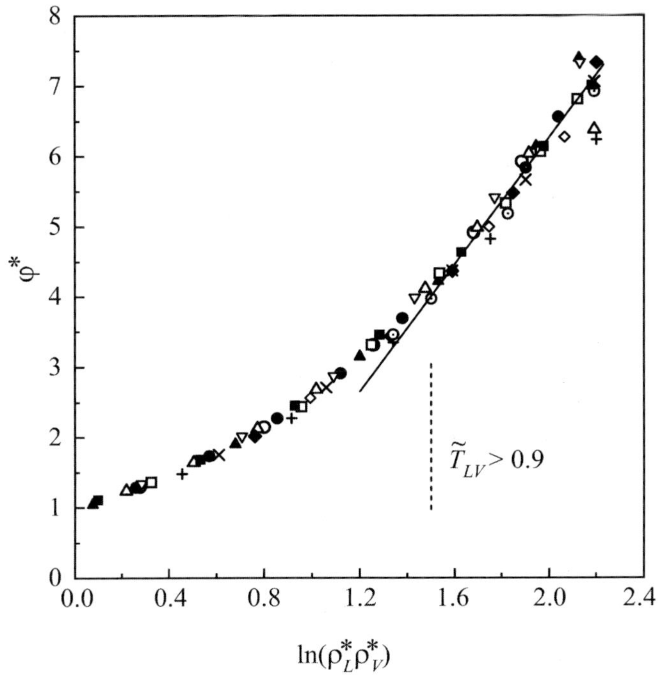

Figure 2.16: Dependence of the quantities φ^* along the saturation curve near the critical point on the product $\rho^*_L\rho^*_V$ for the same group of substances as presented in Fig. 2.12. The full curve is drawn according to Eq. (2.41)

The correlation, demonstrated in Fig. 2.17, can be expressed by the simple relationship

$$\widetilde{\sigma}_{LV} = \left(\frac{\Delta\widetilde{h}_{LV}}{\widetilde{v}_L}\right)^g . \tag{2.47}$$

The average value of the exponent g in Eq. (2.47) for the various substances analyzed is $g = 2.15$. Eq. (2.47) satisfactorily describes the experimental data for temperatures $\widetilde{T}_{LV}$ in the range from 0.6 to 1 and allows us to estimate the surface tension of the liquid by known values of Δh_{LV} and v_L. The mean-square deviation of the experimental data points from the approximation Eq. (2.47) does not exceed 0.5%. The established correlation is a particular expression of the law of corresponding states which does not contain individual parameters of the substances and can be considered as a zeroth-order approximation for the description of the thermodynamic similarity of substances of different nature in variables as introduced with Eq. (2.46).

Let us consider also for various substances the dependence of σ^*_{LV} on the dimensionless variable $(\Delta h_{LV}/v_L p_c)$. Figure 2.18 shows on a logarithmic scale the correlation between these parameters at a reduced temperature $\widetilde{T}_{LV} = 0.6$. Table 2.6 gives the numerical values of the dimensionless variables. As follows from the figure the dependence between the quantities $\lg(\sigma^*_{LV})_{0.6}$ and $\lg(\Delta h_{LV}/v_L p_c)_{0.6}$ is close to a linear one. It has been found that the linearity

Table 2.5: Comparison of the experimental data for φ^* [37] and the data calculated by using Eqs. (2.40) and (2.41). The data in this table are compiled for xenon.

T, K	p, MPa	$\varphi^*_{(exp)}$	$\varphi^*_{(calc)}$	$\dfrac{\varphi^*_{(exp)} - \varphi^*_{(calc)}}{\varphi^*_{(exp)}}$	Equation
161.36	0.0816	0.188	0.190	−0.011	Eq. (2.40)
180	0.2218	0.443	0.443	0.0	
190	0.3480	0.650	0.647	0.004	
200	0.5212	0.914	0.910	0.004	
210	0.7504	1.242	1.236	0.005	
220	1.045	1.641	1.629	0.007	
230	1.416	2.118	2.109	0.004	
240	1.872	2.679	2.673	0.002	
250	2.425	3.350	3.357	−0.002	
260	3.087	4.113	4.044	0.016	Eq. (2.41)
265	3.462	4.537	4.553	−0.004	
270	3.869	4.984	5.048	−0.013	
275	4.310	5.497	5.539	−0.008	
280	4.786	6.032	6.040	−0.001	
285	5.229	6.600	6.540	0.009	

and the slope of this dependence are preserved also at other temperatures up to the critical point ($\widetilde{T}_{LV} = 1$). This result implies that a power dependence can be used to approximate the data, i.e., a function of the form

$$\sigma^*_{LV} = N\left(\frac{\Delta h_{LV}}{v_L p_c}\right)^t . \tag{2.48}$$

The exponent t turns out to be of the order $t = 0.55$. The factor N is a function of temperature with the limiting behavior $N(\widetilde{T}_{LV}) \rightarrow 0$ for $\widetilde{T}_{LV} \rightarrow 1$. For the given temperature, the dimensionless complex

$$N = \sigma^*_{LV}\left(\frac{v_L p_c}{\Delta h_{LV}}\right)^t = \sigma^*_{LV}\left(\frac{p_c}{T_c}\frac{v_L}{\widetilde{T}_{LV}\Delta s_{LV}}\right)^t \tag{2.49}$$

characterizes the similarity of substances with respect to liquid–vapor phase transitions.

Table 2.7 shows the values of N in Eq. (2.48), calculated using experimental data for several temperatures $\widetilde{T}_{LV}$. The calculations were made using the least-squares method. The variation of N along the phase equilibrium line is described by the quadratic dependence

$$1 - \widetilde{T}_{LV} = aN + bN^2 \tag{2.50}$$

with values of the coefficients of $a = 0.870$ and $b = 0.357$.

Tables 2.8 and 2.9 show the values of the surface tension in the temperature range from the triple point to the critical point calculated using Eqs. (2.48) and (2.50). As an example, argon

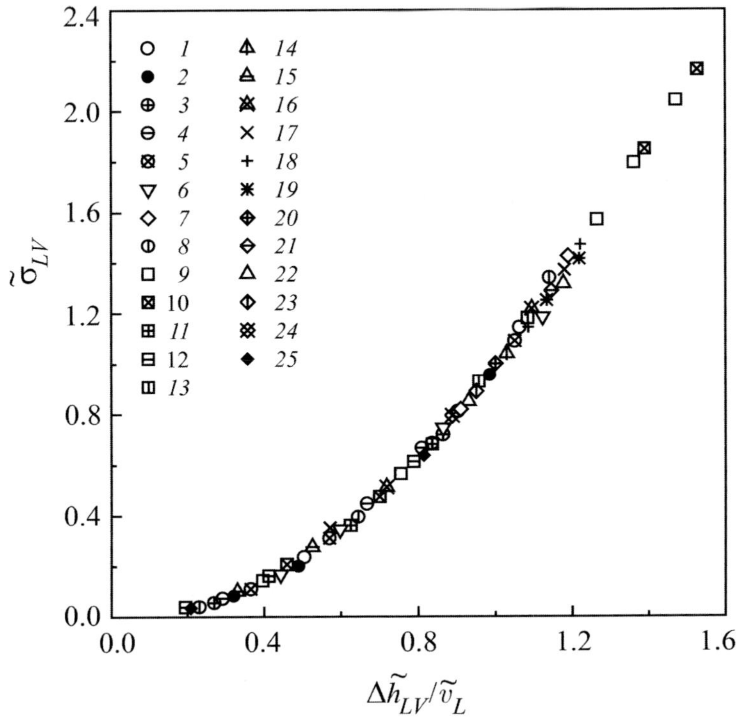

Figure 2.17: Correlation between the dimensionless quantities $\widetilde{\sigma}_{LV}$ and $\Delta\widetilde{h}_{LV}/\widetilde{v}_L$ for normal liquids. The numbers correspond to the numbering of the different substances as given in Table 2.6.

and hexane are considered. The comparison with experimental data gives an indication of the good precision of the established correlation Eq. (2.48). The deviation from the experimental data does not exceed 1% for argon and 3% for hexane. Similarly to Eq. (2.47), Eq. (2.48) does not contain individual parameters and can be used for the calculation of the surface tension using the data for the heat of evaporation and for the density of the liquid at phase equilibrium. The mean-square deviation from the approximative correlation of Eq.(2.48) of the data for 25 normal liquids (Table 2.6) does not exceed 0.5%. This result is comparable to the net accuracy of the experimental measurements of the thermodynamic parameters in Eq. (2.48).

2.4.4 One-Parameter Correlation for the Heat of Evaporation of Nonassociated Liquids

The consideration of the heat of evaporation per unit volume of a liquid phase in the transformation point has been found effective for studying the thermodynamic similarity of substances with respect to the liquid–vapor phase transition. This approach allowed us to establish the law of corresponding states, Eq. (2.48), for the behavior of the surface energy along the equilibrium coexistence curve. To establish the temperature dependence of the dimensionless

Table 2.6: Main parameters defining the thermodynamic similarity of substances on the liquid–gas equilibrium curve: (1) argon, (2) krypton, (3) xenon, (4) nitrogen, (5) oxygen, (6) carbon monoxide, (7) diborane, (8) methane, (9) ethane, (10) propane, (11) butane, (12) isobutane, (13) pentane, (14) isopentane, (15) hexane, (16) heptane, (17) octane, (18) nonane, (19) decane, (20) cyclopentane, (21) cyclohexane, (22) benzene, (23) toluene, (24) monofluorotrichloromethane, (25) hexafluorobenzene.

N	T_c K	p_c MPa	σ_{LV} at $\widetilde{T}=0.6$ mJ/m^2	σ^*_{LV} at $\widetilde{T}=0.6$	$\frac{\Delta h_{LV}}{v_L}$ $\widetilde{T}=0.6$ 10^3kJ/m^3	$\frac{\Delta h_{LV}}{v_L p_c}$ $\widetilde{T}=0.6$	A	L	Refs.
1	150.7	4.86	11.75 [47]	3.192	217.7	44.59	4	1.879	[37]
2	209.4	5.49	14.28 [47]	3.221	248.7	45.30	4	1.887	[37]
3	289.7	5.83	16.87 [47]	3.281	269.3	46.19	4.1	1.894	[37]
4	126.3	3.40	9.20 [47]	3.381	165.5	48.69	3.5	1.906	[39]
5	154.7	5.09	12.50 [47]	3.281	235.3	46.24	3.68	1.893	[38]
6	132.9	3.50	9.95 [47]	3.532	174.4	49.77	3.4	1.919	[40]
7	289.9	4.00	14.37 [40]	3.589	230.3	57.54	2.05	1.991	[40]
8	190.6	4.60	12.36 [47]	3.243	211.1	45.90	3.95	1.891	[42]
9	305.4	4.88	16.31 [47]	3.508	265.5	54.41	2.89	1.968	[54]
10	369.8	4.26	16.66 [47]	3.681	257.8	60.51	2.39	2.013	[55]
11	425.2	3.80	16.90 [47]	3.846	245.9	64.70	2.03	2.040	[40]
12	408.1	3.65	16.40 [47]	3.864	232.0	63.56	2.16	2.013	[40]
13	469.6	3.37	17.30 [40]	4.130	242.8	72.05	1.69	2.086	[40]
14	461.0	3.33	16.83 [40]	4.027	229.8	67.92	1.89	2.061	[40]
15	507.9	3.03	17.26 [40]	4.365	234.7	74.13	1.42	2.111	[40]
16	540.2	2.74	17.32 [40]	4.529	228.7	83.37	1.20	2.140	[40]
17	569.4	2.49	17.05 [40]	4.667	225.5	90.57	1.03	2.174	[40]
18	595.2	2.28	17.01 [40]	4.842	222.1	96.61	0.87	2.220	[40]
19	619.2	2.10	16.89 [40]	5.035	214.8	102.3	0.75	2.252	[40]
20	511.9	4.51	20.64 [40]	3.908	294.5	64.42	2.06	2.047	[40]
21	553.1	4.03	20.83 [40]	4.036	276.1	67.92	1.90	2.063	[40]
22	562.6	4.92	23.09 [40]	4.036	335.8	68.53	1.92	2.061	[40]
23	594.0	4.05	21.61 [40]	4.178	308.1	74.99	1.64	2.108	[40]
24	471.2	4.41	19.65 [52]	3.908	283.6	64.27	2.10	2.033	[40]
25	516.8	3.26	19.98 [53]	4.699	288.6	88.51	1.12	2.162	[53]

Table 2.7: Change of the complex $N(\widetilde{T}_{LV})$ along the phase-equilibrium curve.

$\widetilde{T}_{LV}$	0.5	0.6	0.7	0.8	0.9	0.95	1.0
$N(\widetilde{T}_{LV})$	0.481	0.395	0.308	0.211	0.111	0.057	0.0

variable $\Delta h_{LV}/(v_L p_c)$ one can choose the one-parameter relationship

$$\frac{\Delta h_{LV}}{v_L p_c} = f(A, \widetilde{T}_{LV}) , \tag{2.51}$$

where A is a thermodynamic similarity parameter defined by Eq. (2.10). Employing logarithmic coordinates $\lg\left(\Delta h_{LV}/v_L p_c\right)$ and $\lg\left(1-\widetilde{T}_{LV}\right)$ for plotting the experimental data for

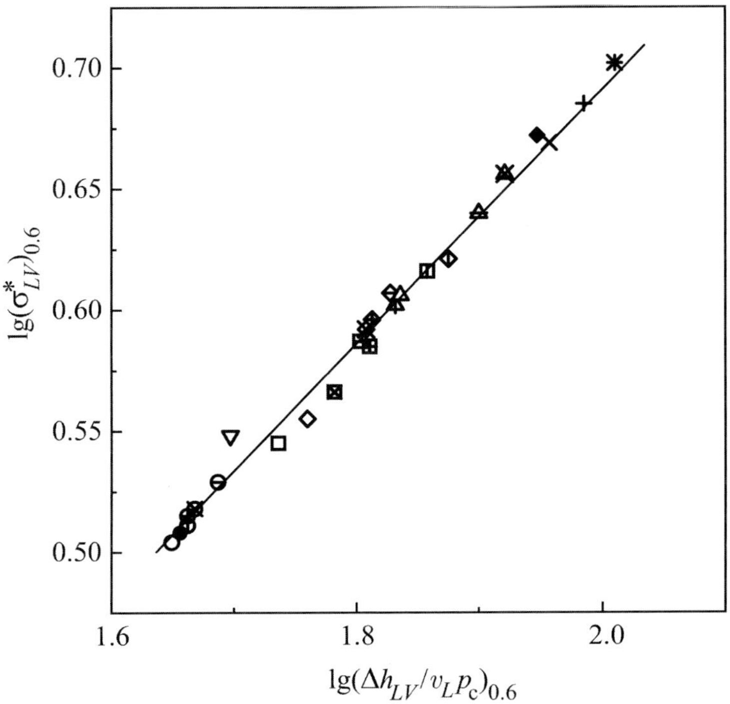

Figure 2.18: Correlation between the dimensionless quantities $\lg(\widetilde{\sigma}_{LV})_{\widetilde{T}=0.6}$ and $\lg(\Delta h_{LV}/v_L p_c)_{\widetilde{T}=0.6}$ for normal liquids. The symbols correspond to the same numbering of the different substances as employed in Fig. 2.17 and Table 2.6.

various substances, we get a linear dependence

$$\lg\left(\frac{\Delta h_{LV}}{v_L p_c}\right) = q\lg(1-\widetilde{T}_{LV}) + L \tag{2.52}$$

with the angular coefficient q, where L is an individual constant. An example of such a plot for argon and benzene is shown in Fig. 2.19. The mean value of the factor q for simple substances like argon is 0.575.

The correlation expressed by Eq. (2.52) means that the function $f\left(A, \widetilde{T}_{LV}\right)$ can be represented as a product

$$f\left(\widetilde{T}_{LV}, A\right) = f_1(A)\, f_2\left(\widetilde{T}_{LV}\right) = f_1(A)\left(1-\widetilde{T}_{LV}\right)^q . \tag{2.53}$$

In this case, the parameter L in Eq. (2.52) equals $\lg[f_1(A)]$.

Figure 2.20 shows the dependence of the parameter L on $\lg A$ for a wide class of substances. The values of L in Eq. (2.52) were calculated using experimental data assigning to q the value $q = 0.575$. Critical parameters of substances used for the calculation of dimensionless variables, the values of the constant L and of the similarity parameter A from [21] are

Table 2.8: Comparison of experimental data for the surface tension $\sigma_{LV(exp)}$ of argon [47] with the values $\sigma_{LV(calc)}$ calculated with Eq. (2.48).

T, K	$\widetilde{T}_{LV}$	$\sigma_{LV(exp)}$, mJ/m^2	$\sigma_{LV(calc)}$, mJ/m^2	$\frac{\sigma_{LV(exp)} - \sigma_{LV(calc)}}{\sigma_{LV(exp)}}$
84	0.557	13.39	13.23	0.012
90	0.597	11.89	11.83	0.005
94	0.624	10.94	10.89	0.004
96	0.637	10.47	10.45	0.002
100	0.664	9.54	9.52	0.003
106	0.703	8.18	8.18	0
112	0.743	6.84	6.86	−0.003
118	0.783	5.55	5.56	−0.001
124	0.823	4.32	4.32	0
130	0.863	3.14	3.13	0.003
136	0.902	2.04	2.04	0
142	0.942	1.055	1.047	0.008
150.7	1	0	0	–

Table 2.9: Comparison of experimental data for the surface tension $\sigma_{LV(exp)}$ of hexane [40] with the values $\sigma_{LV(calc)}$ calculated by using Eq. (2.48).

T, K	$\widetilde{T}_{LV}$	$\sigma_{LV(exp)}$, mJ/m^2	$\sigma_{LV(calc)}$, mJ/m^2	$\frac{\sigma_{LV(exp)} - \sigma_{LV(calc)}}{\sigma_{LV(exp)}}$
273.2	0.538	20.56	20.62	−0.003
303.2	0.598	17.40	17.44	−0.002
333.2	0.657	14.23	14.35	−0.008
363.2	0.716	11.22	11.36	−0.013
393.2	0.775	8.34	8.44	−0.012
423.2	0.834	5.62	5.70	−0.014
453.2	0.893	3.16	3.25	−0.027
473.2	0.933	1.74	1.79	−0.026
493.2	0.972	0.58	0.58	0
507.4	1	0	0	–

given in Table 2.6. The dependence shown in Fig. 2.20 is close to a linear one and allows us to determine the form of the function $f_1(A)$. Processing the data by using the least squares method one gets the correlation

$$f_1(A) = 155A^{-0.5} \,. \tag{2.54}$$

Then the generic correlation Eq. (2.51) takes the following form

$$\frac{\Delta h_{LV}}{v_L p_c} = 155\, A^{-0.5}(1 - \widetilde{T}_{LV})^{0.575} \,. \tag{2.55}$$

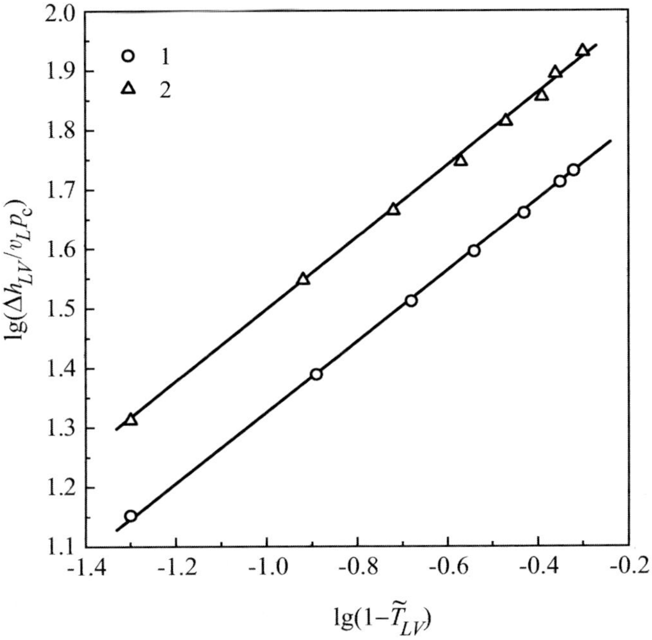

Figure 2.19: Correlation between the quantities $\lg(\Delta h_{LV}/v_l p_c)$ and $\lg\left(1 - \widetilde{T}_{LV}\right)$ for argon (1) and benzene (2).

Table 2.10: Comparison of experimental data for the heat of evaporation $\Delta h_{LV(exp)}$ of carbon monoxide [40] with the calculated ones $\Delta h_{LV(calc)}$ by using Eq. (2.55).

T, K	$\widetilde{T}_{LV}$	$\Delta h_{LV(exp)}$, kJ/kg	$\Delta h_{LV(calc)}$, kJ/kg	$\dfrac{\Delta h_{LV(exp)} - \Delta h_{LV(calc)}}{\Delta h_{LV(exp)}}$
70	0.526	233.3	228.2	0.022
80	0.602	218.5	217.1	0.006
94	0.707	200.0	204.0	−0.020
100	0.752	182.9	186.6	−0.020
110	0.828	160.3	164.5	−0.026
120	0.903	133.6	133.0	0.004
125	0.941	102.5	109.6	−0.070
129.85	0.977	71.1	74.5	−0.048
132.9	1	0	0	–

Figure 2.21 shows the dependence of $\Delta h_{LV}/v_L p_c$ on $f\left(A, \widetilde{T}_{LV}\right)$ in the range from the triple point to temperatures close to the critical one for various substances in logarithmic coordinates. The mean-square deviation of this point array from the correlation Eq. (2.55) does not exceed 1%. This correlation can be used to calculate the heat of evaporation Δh_{LV} by known values of v_L along the saturation line.

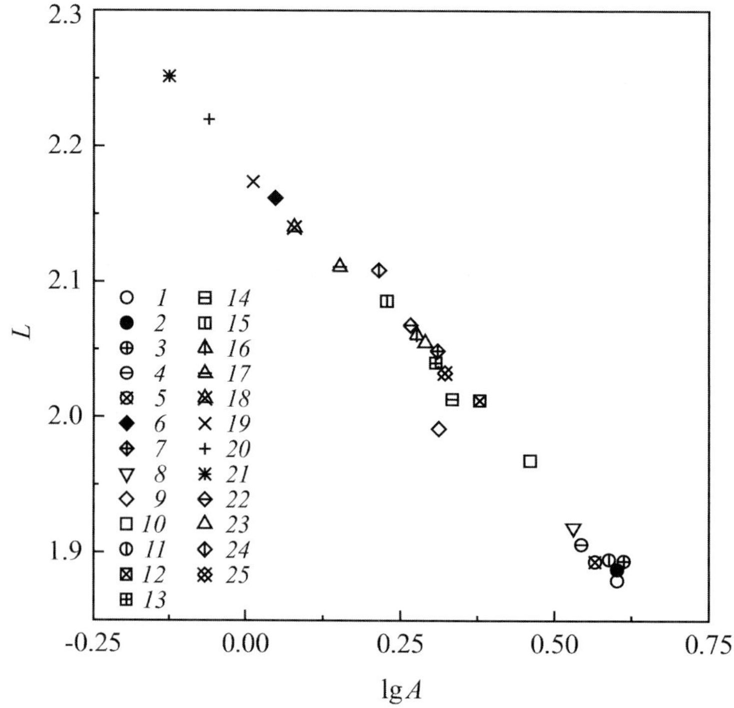

Figure 2.20: Dependence of the values of the constant L on the parameter A for normal liquids. The numbering of the different points corresponds to that given in Table 2.6.

Table 2.11: Comparison of experimental data for the heat of evaporation $\Delta h_{LV(exp)}$ of benzene [40] with the calculated ones $\Delta h_{LV(calc)}$ by using Eq. (2.55).

T, K	$\widetilde{T}_{LV}$	$\Delta h_{LV(exp)}$, kJ/kg	$\Delta h_{LV(calc)}$, kJ/kg	$\frac{\Delta h_{LV(exp)} - \Delta h_{LV(calc)}}{\Delta h_{LV(exp)}}$
283.2	0.504	442.3	412.0	0.069
323.2	0.575	415.5	395.8	0.047
353.2	0.628	394.6	380.9	0.035
423.2	0.753	337.7	335.8	0.006
473.2	0.842	286.8	287.1	−0.001
523.2	0.931	208.0	210.6	−0.012
553.2	0.984	115.3	111.8	0.030
562.1	1	0	0	–

Tables 2.10 and 2.11 using the examples of carbon monoxide and benzene show the values of Δh_{LV} in the temperature range from the triple point to the critical point calculated using Eq. (2.55). The comparison with experimental data gives a strong indication of the good precision of the established correlation.

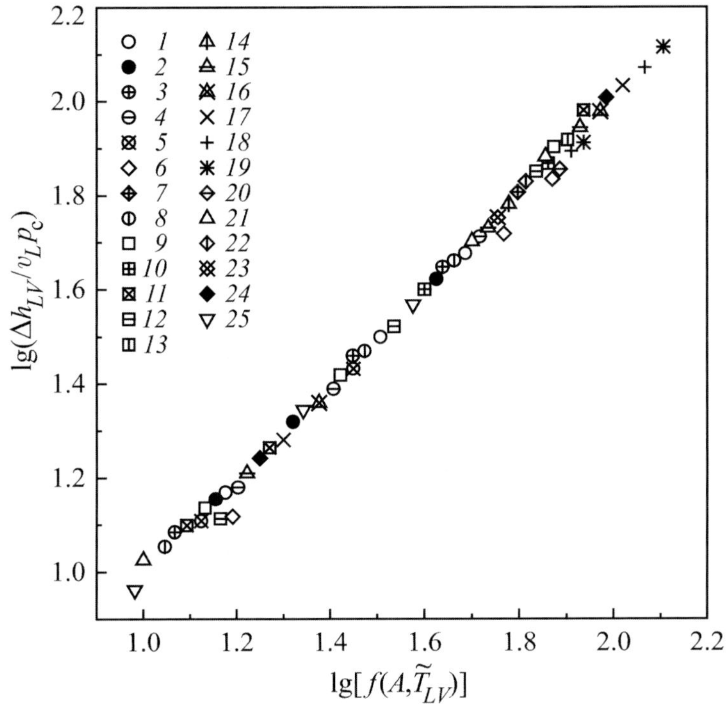

Figure 2.21: Correlation between the quantities $\lg(\Delta h_{LV}/v_l p_c)$ and $\lg\left[f\left(A, \widetilde{T}_{LV}\right)\right]$ for normal liquids. The specification of the different point corresponds to the numbering as given in the caption to Table 2.6.

Although being of interest itself, the consideration of the liquid–vapor phase transition also serves as an introduction to the methodology of thermodynamic similarity and prepares the reader for the next chapter, where the crystal–liquid transition will be considered using an analogous approach. Having this further application in mind, the appropriate material has been chosen as outlined in the present chapter. The existence of the critical point on the fluid phase equilibrium coexistence curve ensures the possibility to generalize the data for various substances in the form of similarity correlations. Some new correlations were introduced, which reflect the universal character of relations between thermodynamic parameters, for instance, between the saturation pressure and the product of orthobaric densities of liquid and gas. In the next chapter, an attempt is made to derive similar relationships for liquid–crystal coexistence. In order to proceed in such a way, immediately the question arises whether there exists a critical point along the liquid–crystal coexistence curve and, if not, which state may serve as the reference state to introduce appropriate scaling parameters.

3 Crystal–Liquid Phase Transitions

3.1 The Behavior of the Crystal–Liquid Equilibrium Curve at High Pressures

The crystal–liquid phase transition is a typical form of phase transition of the first kind. In order to melt a crystal, heat has to be supplied to the crystal phase. At constant pressure, the melting then proceeds at constant temperature. The transition is accompanied by changes of volume and entropy of the substance. The melting curve separates two metastable states: the supercooled liquid and the superheated crystal. Since melting is accompanied by supply of heat to the crystal, the entropy of the substance is increased and $\Delta s_{SL} > 0$ holds. Moreover, for most substances, we can observe the "normal" type of the melting curve which corresponds to positive values of the change of the volume, Δv_{SL}, and positive values of the derivative, dp/dT_{SL}. Such a type of equilibrium curve means that the crystal phase is more ordered and dense than the melt. At the same time, there are substances (like gallium, bismuth, and water) having a negative slope of the derivative $dp/dT_{SL} < 0$. In such cases, the density of the liquid is higher than the density of the crystal ($\Delta v_{SL} < 0$) at positive values of Δs_{SL}. Such substances, which may be denoted as "anomalously" melting systems, have, as a rule, a loose crystal structure with a small coordination number. The melting of such substances results in the formation of a disordered structure with a more dense packing of the particles. This property is the origin of the decrease of the volume and, consequently, the negative slope of the equilibrium coexistence curve.

For some substances (like rubidium, cesium, and barium), the derivative dT_{SL}/dp along the melting curve becomes equal to zero with an increase of pressure and then becomes negative. The melting curves of such substances have a temperature maximum. According to Eq. (1.2), the crystal and the melt have the same density at the maximum, i.e., $\Delta v_{SL} = 0$. This result implies that a variation of pressure can lead to a change of the sign of the difference, Δv_{SL}. This inversion indicates that the compressibility of the liquid is higher than that of the crystal [56]. Experiments show that, with an increase of the pressure along the "anomalous" segment of the coexistence curve, such substances undergo a polymorphic transition inside the crystal phase. After this transition is completed, the slope of the melting curve becomes positive again.

An overview of some theoretical approaches, explaining the phenomenon of the maximum of the melting curves, can be found in Ref. [57]. In Ref. [58], a qualitative analysis of the form of the melting curve is given in (p, T)-coordinates for the extremal conditions of dense nuclear matter. The author concludes that the melting curve bends with the increase of the pressure

Crystal–Liquid–Gas Phase Transitions and Thermodynamic Similarity. Vladimir P. Skripov and Mars Z. Faizullin

ISBN: 3-527-40576-3

and the derivative (dp/dT_{SL}) changes its sign from plus to minus. As a general rule, this inversion takes place at a density exceeding the density of common molecular systems by several orders of magnitude. The detailed analysis of this particular case is out of the scope of the present book.

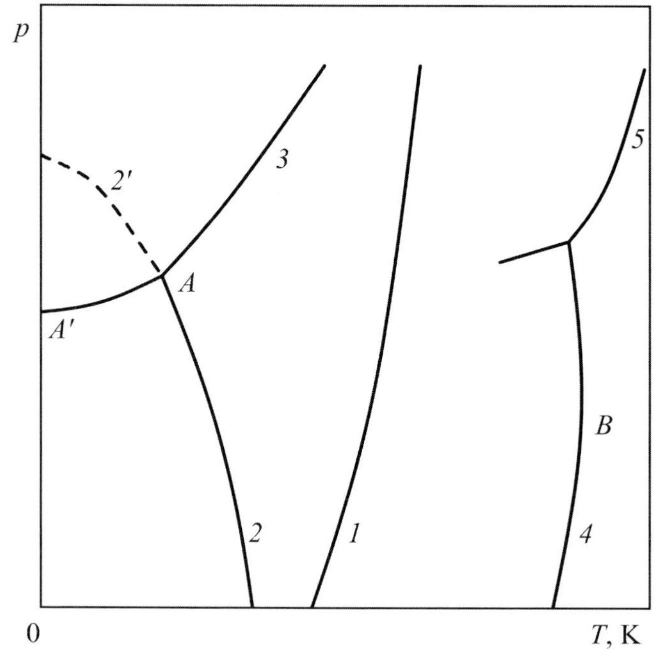

Figure 3.1: Characteristic types of liquid–crystal equilibrium coexistence curves of one-component systems (see text).

Summarizing the above results, we conclude that the existing experimental information about the melting of substances at high pressures allows us to distinguish three types of phase diagrams of one-component systems in pressure–temperature coordinates. There are phase diagrams with an infinitely increasing melting curve, phase diagrams with "anomalously" behaving melting curves and phase diagrams having a temperature maximum along the coexistence curve. These different types of behavior of the melting curves are shown in Fig. 3.1. Here 1 specifies the melting curve of "normal" type without polymorphic transformations inside the crystal phase, and 2 and 3 are the melting curves with a triple point A corresponding to an equilibrium coexistence of liquid and two crystal phases. The phase of low pressure is stable below the curve AA′ and melts with a negative slope of the melting curve, $(dp/dT_{SL}) < 0$. The dashed segment $2'$ is plotted as the metastable continuation of the melting curve 2. The characteristic features of the coexistence curves specified by the number 4 are the point B of the temperature maximum of the melting curve, the triple point and the transition to the high-pressure phase with a positive slope of the melting curve, indicated by curve 5. One may expect that for high pressures, when the packing of the particles in all substances is tight, the behavior of the melting curves will have generic common features for different substances regardless of their nature.

The question about the behavior of the melting curves at high pressures is still a subject of scientific discussions. In contrast to the liquid–gas equilibrium curve, the kind of high-temperature asymptotic behavior of the melting curve is not so evident. Great efforts have been made in order to determine the critical point on the melting curve, above which a continuous transition from one phase to the other is possible. The decrease of the volume differences, Δv_{SL}, and of the melting entropy, Δs_{SL}, with an increase of temperature allows one to expect that the values Δv_{SL} and Δs_{SL} will possibly tend to zero. Extrapolations of the Δv_{SL} and Δs_{SL} dependencies as a function of pressure and temperature were used to estimate the parameters of the critical point on the melting curve [59]. For example, for potassium the values of critical pressure, p_c, and critical temperature, T_c,

$$p_c = 2196\,\text{MPa}\,, \qquad T_c = 498\,\text{K} \tag{3.1}$$

were estimated in that way. However, subsequently performed experimental investigations have significantly exceeded the mentioned values of pressure and temperature and did not exhibit any singularities on the melting curve of potassium [60].

The possibility of the existence of a critical point on the melting curve is indicated by the well-known theory of melting due to Lennard-Jones and Devonshire [61]. According to this theory, above the temperature

$$T_c = 1.1\frac{\varphi_0}{k_B} = 1.57\,T_m \tag{3.2}$$

the isotherms loose their s-shaped form similar to the liquid isotherms at the critical point of liquid–gas equilibrium. Here φ_0 is the energetic parameter directly related to the atomic interaction forces, k_B is the Boltzmann constant, and T_m is the melting temperature at atmospheric pressure. The values of the critical point parameters, calculated using the Lennard-Jones–Devonshire model are located significantly below the values of temperature and pressure where the course of the melting curve can be measured.

Analyzing the presently available experimental data about the melting of substances, one can draw the conclusion that the melting temperature increases infinitely with the increase of pressure and that a critical point on the melt–crystal equilibrium curve is absent. The absence of a critical point singularity is deeply connected with the essential discontinuity inherent in the properties of crystal and liquid. The properties of each of these condensed phases are described by its own equation of state. This property reflects the fact that there exist two qualitatively different modes of dense packing of atoms, the regular and the irregular packing modes, characterized by a different symmetry.

Symmetry-based concepts for the substantiation of the absence of an equilibrium critical point on the liquid–solid coexistence curve have been formulated first by Landau [2, 62]. According to his argumentation, in the transition from the solid to the liquid some symmetry elements characteristic to the crystal phase disappear. However, symmetry elements cannot disappear continuously. They are either present, or absent. Consequently, a continuous transition between different phases, which is characterized by a change of the symmetry of the system under consideration, is also impossible. Note that the account of the existence of the thermal motion of the atoms or molecules constituting the respective substances leads to the conclusion that the symmetry-based approach to the understanding of essential features of the melt–crystal transition is connected with some characteristic time scale, the averaging time of the spatial distribution of the atoms or molecules of the system under consideration.

3.2 Experimental Methods of Investigation of Melting of Substances at High Pressure

Systematic experimental investigations of the melting phenomenon started with the work of Tammann at the end of the 19th century. The pressure in his experiments reached values up to 300 MPa. Tammann was the first who used the method of discrete volume change observation at the melting point (shifting piston method). This method consists in the following. At a given temperature, the liquid under investigation is transferred into the solid phase by increasing the pressure in the piezometer. Then, by slowly releasing the pressure, the shift of the location of the piston is observed. At the melting point, a jump-like shift of the piston occurs due to the change of the volume of the sample in the transition from the solid to the liquid phase. The advantage of this method is that it allows one to measure not only the dependence of the melting temperature on pressure, but also the volume effect of melting and then to calculate the changes of other thermodynamic functions along the melting curve using the obtained data.

Later the method of the shifting piston was further developed in the experiments performed by Bridgman. He is also the author of the first analyses where the melting of metals at high pressures were investigated using a jump of electric resistance during the phase transition. Bridgman's achievements in advancing the technology of high pressures allowed him to extend the range of pressure in the experiments up to several GPa. The results of his investigations of melting of various substances at high pressures are given in his monographs [63, 64]. They also contain an overview of the most significant investigations in this area by other authors up to 1945.

The measurement of melting curves of molecular substances was often carried out using the capillary plugging method proposed by Kamerlingh-Onnes in 1926. A detailed description of this method can be found in Ref. [65]. The investigated substance is brought at some given pressure into two vessels connected by a capillary. With the cooling of the capillary, the liquid contained in it becomes solid and blocks the capillary. In the case of an increase or decrease of the pressure in one of the vessels the pressure in the other one remains constant. Melting occurs with the gradual heating of the capillary when the difference of pressures in the vessels is about 0.1 MPa. The melting process is reflected then by the approach of the pressure in both vessels to the same value.

In Refs. [66, 67], a description of the experimental investigations of melting of a group of fluoroorganic compounds in a range of pressures from 0.1 to 230 MPa is given. The melting curves have been measured using the heat effect of melting and the change of dielectric permittivity of the substance in transition from the crystalline to the liquid phase. Figure 3.2 shows the high-pressure chamber with a test cell containing the capacitance melting sensor and the copper–constantan thermocouple. The sensor itself represents a cylindrical capacitor with a coating made from steel thin-wall pipes. The chamber is dipped into a liquid thermostat which allows the temperature to be changed in the interval from 180 K to 350 K and to keep

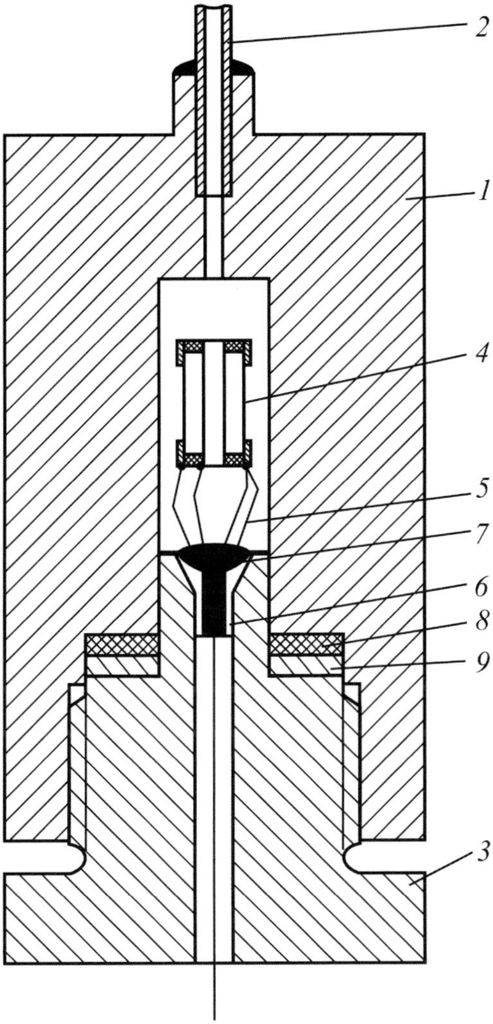

Figure 3.2: High-pressure chamber: (1) steel bomb, (2) steel capillary, (3) screw, (4) cylindrical capacitance, (5) thermocouple, (6) cone of electric cabels, (7) densified part consisting of epoxy resin, (8) teflon gasket, (9) brass gasket.

it with a precision of ± 0.01 K. The investigated substance is transferred into the solid phase by an increase of pressure at a constant temperature in a thermostat. When the temperature is established, the pressure is decreased by a rate of 20–30 kPa/s.

Figure 3.3 shows the curves for monofluorobenzene at $T = 255.74$ K obtained by a thermocouple and capacitance sensor resulting from the processes in the substance caused by the decrease of pressure. Segments 1 and 3, respectively, of the curves correspond to the solid and liquid phases. As shown, the process of melting is accompanied by an absorption of heat

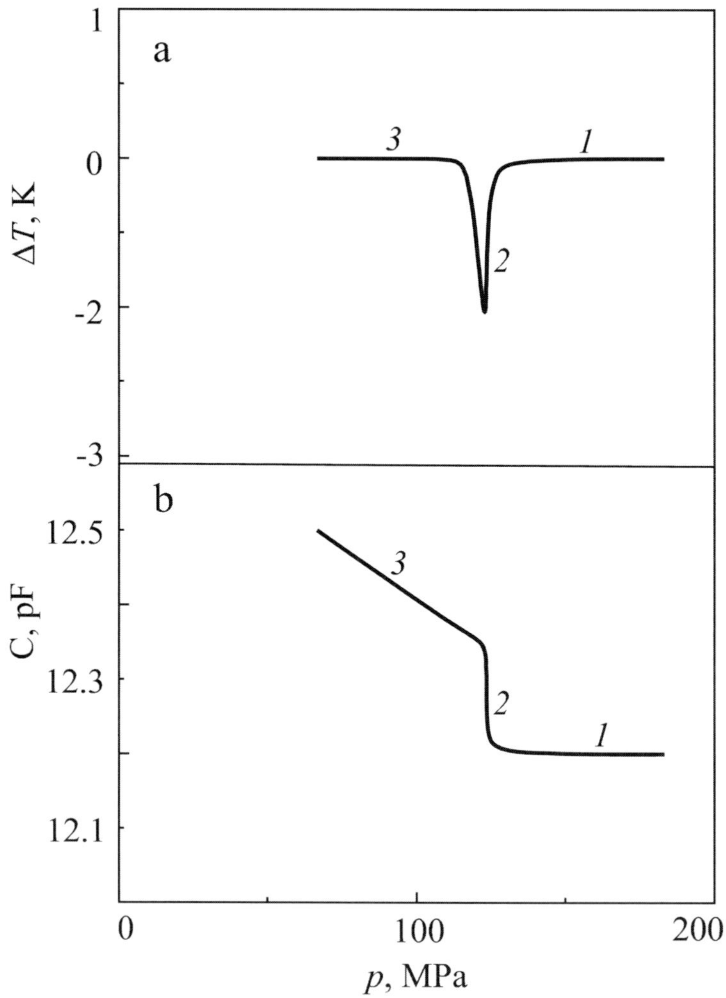

Figure 3.3: Results of measurements by a thermocouple (a) and a capacitance sensor (b) obtained in the melting of monofluorobenzene at $T = 255.74$ K.

(registered by the thermocouple) and the jump in the capacity of the sensor connected with the change of the dielectric permittivity of the investigated substance in the phase transformation (part 2 of the curves in Fig. 3.3).

The measurement results for the melting curves of four fluoroorganic substances in a pressure range from 0.1 to 230 MPa are given in Fig. 3.4. The dependences of melting pressure on temperature were approximated using the Simon equation [68]

$$\frac{p}{p_*} = \left(\frac{T}{T_0}\right)^c - 1\,, \tag{3.3}$$

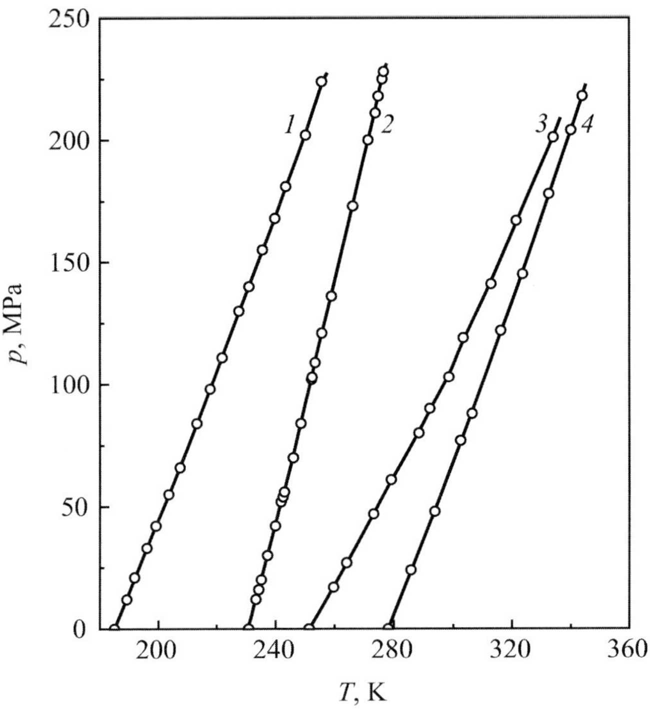

Figure 3.4: Dependence of the melting temperature on pressure for the following substances: (1) perfluorohexane, (2) monofluorobenzene, (3) perfluorooctane, and (4) hexafluorobenzene.

where p_* and c are empirical constants and the temperature T_0 corresponds to the melting temperature at a pressure $p = 0$. In practice, T_0 is chosen equal to the melting temperature at atmospheric pressure. Making such replacement, the corresponding correction to T_0 for the investigated substances is ≈ 0.03 K, which does not exceed the error in the experimental measurements of the melting temperature.

In Table 3.1, the following parameter values of the Simon equation (Eq. (3.3)) for the substances shown in Fig. 3.4 are given: temperature T_0, pressure p_*, exponent c, mean square errors of the determination of these parameters, $\sigma(p_*)$ and $\sigma(c)$, and also the values of the initial slope of the melting curves, $(dp/dT_{SL})_{T_0}$, calculated by using the Simon approximation.

In the last few decades, the possibility was realized to make experimental investigations on crystal–melt phase transitions at pressures above 10^4 MPa and at temperatures exceeding 1000 K. The application of the above discussed traditional methods of melting temperature measurement using thermocouples is impossible under such conditions since the material of the thermocouple would also melt. In Ref. [69], the melting curve of graphite up to 11000 MPa is determined in experiments employing the impulse heating of the sample. The melting temperature has been calculated from the capacity of an electric current heating the sample. In Ref. [70] the melting curves of graphite, wolfram and platinum are studied for pressures up

Table 3.1: Parameters of the Simon equation and the slope of the melting curves at a temperature T_0 for monofluorobenzene, hexafluorobenzene, perfluorohexane and perfluorooctane (see text).

Substance	T_0, K	p_*, MPa	$\sigma\,(p_*)$, MPa	c	$\sigma\,(c)$	$(dp/dT_{SL})_{T_0}$, MPa/K
Monofluorobenzene	231.0	601	22	1.79	0.06	4.7
Hexafluorobenzene	278.1	454	28	1.85	0.10	3.0
Perfluorohexane	185.0	340	13	1.56	0.05	2.9
Perfluorooctane	251.4	222	15	2.27	0.12	2.0

to 6000 MPa using an optical method. The temperature of the sample was measured relative to the intensities of two spectral curves of the continuous emission spectrum of the investigated substance according to the laws of thermal emission. The sample is heated at a given pressure by an electric current. Melting leads to a breakdown of the electric circuit, and the melting temperature was determined at the moment when the intensity of the spectral curves stopped increasing.

The melting of carbon at high pressures using stationary laser heating of a sample has been studied in Ref. [71]. The temperature in the experiment has been measured by using a fast pyrometer. The phase transformation has been registered by a horizontal segment of the heating thermogram (in the curve giving the dependence of sample temperature on time) at a fixed intensity of laser emission intensity and fixed pressure in the experimental chamber. In Ref. [72], the melting curves of some metals have been determined from the equations of state of the solid and liquid phases. These equations of state were developed using experimental data of impact compression. Along with the melting curve, this method allows one to determine the change of volume and other thermodynamic functions along the equilibrium coexistence curve for pressures up to $\sim 10^5$–10^6 MPa. In recent years, experimental investigations under pressure have involved diamond anvils, which allow one to observe the melting of substances in conditions of static compression at pressures up to $\sim 10^5$ MPa. For example, in Ref. [73] the melting curves were measured in such an experimental cell by laser heating of samples of some alkali halides at pressures up to 37 GPa and temperatures up to 3500 K.

Numerous experimental data on the melting processes of various substances, obtained during more than 100 years of investigations, are given in Refs. [74–77]. Refs. [74,75] contain the (p,T)-dependencies for a wide range of elements with the specification of measurement method and references to original literature. Experimental data for the melting curves of ionic compounds can be found in the review Ref. [76] and in the reference book Ref. [77]. Reference [78] contains the results of processing the melting curves of more than 200 different substances using the Simon approximation. Also references to experimental investigations of melting of these substances are given.

3.3 Application of Similarity Methods for a Description of Melting

The application of the method of thermodynamic similarity for a description of crystal–liquid phase transitions implies a thorough analysis of a large amount of experimental data in order to establish common laws governing the thermodynamic properties in the melting process. Having established such laws, we can extend our knowledge about melting to poorly studied substances or to predict the behavior of thermodynamic properties characterizing the melting process in wider ranges of temperature and pressure. However, severe complications in realizing such kind of program occur.

The obstacles in realizing such an approach for studying melting are due to the diversity and complexity of melting processes which may proceed quite differently for different substances. Along with the evolution of translational disorder there are orientational, configurational and oscillatory melting mechanisms. As a consequence, we have to introduce into the theoretical description the respective contributions to the melting entropy [59]. In Ref. [79], in addition, the notion of electronic disorder is introduced in application to semiconductors metallizing during the transition from the solid to the liquid state. A significant contribution to the change of entropy in the course of the melting process of such substances is caused by the increase of the number of conducting electrons in melting.

Most of the papers dedicated to the investigation of the melting process from the position of thermodynamic similarity are reduced to establishing the correlations between various properties in the triple point or at atmospheric pressure. The first attempts trying to establish universal features of the melting process were directed exclusively to the comparison of the thermodynamic properties characterizing the melting process [59]. It was found that some groups of substances have nearly constant values of the melting entropy, Δs_{SL}, and of the relative volume jump in melting, $(\Delta v_{SL}/v_S)$, where v_S is the specific volume of the crystal phase. An analysis of experimental data allows us to conclude that the behavior of thermodynamic properties in the melting point is rather specific for various substances. But for some groups of substances (e.g., alkali metals, inert gases) having a similar character of the intermolecular potentials and similar crystal structures one can establish similarities in the variation of the thermodynamic properties during the melting process. Similar results can be achieved if we compare the reduced (p, v, T)-values characterizing the melting process and choose as scaling values the parameters of the liquid–gas equilibrium critical point [80]. However, critical parameters are the natural scaling values for the description of liquid–gas phase transitions, but in application to melting the similarity in the reduced values can be considered as useful only for very limited groups of substances.

The analysis of experimental data of thermodynamic properties at the melting point for a large group of simple substances allows one to establish a number of relationships indicating that the properties are indeed correlated. For example, Ref. [81] contains an analysis of the correlation between the melting temperatures, T_0, of some crystals and their enthalpies, h_S, at the melting point. The dependence of the melting temperature on h_S (in kJ/mol) can be represented by the following empiric relationship

$$T_0 = 75.2\left(h_S^{0.8} + 2.660\right) . \tag{3.4}$$

The enthalpy of the crystalline phase, h_S, and the latent heat of the phase transition, the enthalpy of melting, Δh_{SL}, are found to be linearly dependent. For metals, one gets

$$\Delta h_{SL} = 0.29\,(h_S + 2.76) , \tag{3.5}$$

and for nonmetallic substances the relation

$$\Delta h_{SL} = 1.568\,(h_S + 0.765) \tag{3.6}$$

holds. These results imply that the heat required for the melting of the crystal is proportional to the enthalpy of the crystal at the melting point, and that the coefficient of proportionality depends on the kind of bonding of the basic atomic or molecular units in the substance.

Reference [82] describes an attempt to establish a correlation between the crystal structure of solids and the thermodynamic properties at the melting point at atmospheric pressure. For various groups of elements having a common crystal structure the following linear dependencies of the melting entropy Δs_{SL} (in J/(K mol)) on the melting temperature T_0 are proposed

$$\Delta s_{SL} = 6.795 + 8.742 \cdot 10^{-4} T_0 \qquad \text{for bcc crystals} \tag{3.7}$$

$$\Delta s_{SL} = 8.093 + 8.742 \cdot 10^{-4} T_0 \qquad \text{for fcc crystals} \tag{3.8}$$

and

$$\Delta s_{SL} = 9.307 + 8.742 \cdot 10^{-4} T_0 \tag{3.9}$$

for cpa (close-packed atomic) crystals. Examples of some other correlations between thermodynamic properties characterizing melting at atmospheric pressure can be found in Refs. [59, 79].

The application of thermodynamic similarity concepts for a description of melting processes in a wide range of thermodynamic parameters is complicated by the problem of a proper choice of reasonable scaling parameters for each of the variables p, v, and T. The construction of dimensionless thermodynamic parameters can be facilitated by considering the molecular properties of the substance. The general character of intermolecular interaction and the similarity of the crystal structure of some simple substances, mostly having the melting entropy of a positional, translational nature, allows one to establish the similarity in the behavior of these substances in the melting process. The interaction potential parameters can be used to construct dimensionless reduced values of temperature, T^*, pressure, p^*, and volume, v^*, respectively. Indeed, in the following coordinates, $T^* = k_B T/\varepsilon$, $p^* = \sigma^3 p/\varepsilon$, and $v^* = v/N_A\sigma^3$, where ε and σ are the Lennard-Jones potential parameters, and N_A is the Avogadro number, the law of corresponding states holds true for the melting curves of argon, krypton, xenon and, partially, for neon [83]. For the group of alkali metals, the similarity of melting curves is established for sodium, potassium, rubidium and cesium in the framework of the pseudopotential theory [84].

Some empiric and semiempiric relations for the caloric properties of the crystal phase along the melting curve can be considered as a manifestation of the similarity with respect to the melting process. Among them are the Ross rule and entropy criteria [85]. According to the criterion, the following correlation holds true along the melting curve

$$\frac{\varphi(T, v_S) - u_0(v_S)}{N_A \theta} = \text{constant}\,, \tag{3.10}$$

where $u_0\,(v_S)$ is the potential energy of the atoms if they are located at the lattice points, v_S is the specific volume of the crystal along the melting curve, $\varphi\,(T, v_S)$ is the configurational (i.e., excessive in relation to the ideal gas) free energy of the crystal, and θ is the Debye temperature. Similarly, according to the entropy criterion, the configurational energy of a crystal remains constant on the melting curve

$$\frac{s_S - s_{id}}{N_A k_B} = \text{constant} , \tag{3.11}$$

where s_{id} is the entropy of the ideal gas. In a classical quasi-harmonic approximation these correlations are consequences of the Lindeman rule [86], which states that the ratio of root-mean-square deviation of atoms from the equilibrium position in the crystal to the interatomic spacing remains constant along the melting curve. Using the thermal equation of state for the crystal phase these criteria allow us to calculate the melting curve for a given substance. The results of such calculations for some simple substances (argon, sodium [85]) are in good agreement with experiment.

Equations (3.10) and (3.11), as well as the Lindeman rule itself, represent examples of the one-phase approach to the description of the melting process, i.e., the properties of the liquid–solid equilibrium curve are deduced from the knowledge of the properties of only one of the coexisting phases. Such criteria cannot be of universal character, and their application is restricted to substances with a relatively primitive crystal structure. A consideration of the high-temperature asymptotics of the thermodynamic properties along the melting curve for substances with densely packed and similar structures can be found in Refs. [56, 83]. These analyses include also a consideration of the melting thermodynamics of such substances. The experimental data, presently available, allow us to conclude that the behavior of the melting curves of substances like argon remains unchanged up to very high pressures, i.e., the melting temperature increases steadily with the increase of pressure. The discontinuous character of entropy and volume variation indicates that melting retains the attributes of a phase transition of first order for the entire experimentally studied pressure range. The melting entropies, $\Delta s_{SL}/R$, and volume changes, $\Delta v_{SL}/v_S$, rapidly decrease along the first segment of the melting curve and tend to constant values with a further increase of pressure and temperature. Extrapolating the data to the region of high pressures, one obtains for argon the following asymptotic values for the variation of entropy and the relative volume jump at melting [56]

$$\frac{\Delta s_{SL}}{R} \approx 0.9 , \qquad \frac{\Delta v_{SL}}{v_S} \approx 0.03 . \tag{3.12}$$

A computer simulation experiment on the melting process in a system of hard spheres interacting by the law $\varphi(r) = \varepsilon(\sigma/r)^{12}$, where ε and σ are constants with the dimensions of energy and length, has given similar values for entropy and volume jumps on the melting curve [56]. The results of the numerical simulation are

$$\frac{\Delta s_{SL}}{R} \approx 0.89 , \qquad \frac{\Delta v_{SL}}{v_S} \approx 0.038 . \tag{3.13}$$

This coincidence in the values of the respective parameters allows us to conclude that the high-temperature asymptotic behavior of the thermodynamic properties on the melting curve for substances like argon is closely related to the melting of a system of solid spheres and is defined by the repulsive part of the interatomic interaction potential.

The behavior of the quantities $\Delta s_{SL}/R$ and $\Delta v_{SL}/v_S$ along the melting curve for argon and sodium has revealed a dependence close to a linear one between these two parameters [83]. The form of the function $\Delta s_{SL}/R = f(\Delta v_{SL}/v_S)$ is the same for both substances. According to this dependence the limiting (for $\Delta v_{SL}/v_S \to 0$) value of the melting entropy $\Delta s_{SL}/R$ turned out to be close to $\ln 2 \cong 0.7$. In the opinion of the author of Ref. [83], the value $R \ln 2$ can be considered as the lower limit for the melting entropy: it characterizes the minimal "disorder" occuring in the course of the crystal–liquid phase transition.

In Ref. [87], the correlation of this value with the collective entropy is considered. It is noted there that it is still unclear to which degree the experimentally determined asymptotic value of the melting entropy should be found near to $R \ln 2$. The principal character of this question is caused by the fact that the limiting value of the melting entropy, $\Delta s_{SL}/R$, can be considered as a manifestation of configurational disorder in objects of very different physical nature [88–94]. The mentioned approach, which is based on the analysis of the high-temperature asymptotics and the search for universal features in the behavior of the melting entropy at high pressures has been further developed in Refs. [95–97] both for simple substances and for compounds having linear and more complex molecular structures.

The review given above shows that the correlations obtained so far, reflecting the similarity of substances with respect to melting, cover either a narrow range of substances or are related only to atmospheric pressure. As a consequence, they cannot be used as a basis for establishing algorithms for calculating the melting curves of different substances in a wide range of change of the thermodynamic variables. The absence of a critical point on the melting curves leads to the problem of choosing the scaling variables for pressure and, consequently, for temperature. The following considerations change the direction of the search for these scaling parameters by directing the attention not to the high-temperature but to the low-temperature asymptotics of the melting curves. The realization of such a program constitutes the main content of the following sections of this chapter. Various consequences, which can be obtained employing such an approach, are considered, and a comparison of the results with experiment is made.

3.4 The Extension of the Melting Curve into the Range of Negative Pressures and the Scaling of Thermodynamic Parameters

The extension of the phase equilibrium curve into the region of metastable states of the coexisting phases is not forbidden by physical arguments. It is presumed in such an extrapolation that both phases retain internal stability with respect to continuous variations of their state. For crystal–liquid equilibrium, the above statement implies the extension of the melting curve beyond the triple point into the region of negative pressures (to stretched or expanded states of the coexisting phases). Experimentally such an extension of the melting curve has been observed for ice down to $p = -24$ MPa [98]. The condition for phase equilibrium is given by Eq. (1.1) or, in a more expanded form, by

$$u_S + pv_S - Ts_S = u_L + pv_L - Ts_L \ , \tag{3.14}$$

where u is the specific internal energy, s is the specific entropy, v is the specific volume, and the indices S and L refer to the crystal and liquid phases, respectively. This condition can be fulfilled both for positive and negative pressures.

The extension of the melting curve into the region of negative pressures has been employed already for a description of the nucleation process in supercooled melts [6]. For tin, as an example, the formation of nuclei of the crystalline phase in the liquid by homogeneous nucleation is observed at atmospheric pressure at a temperature $T = 384$ K. This result corresponds to a supercooling of the melt of $\Delta T = 121$ K below the equilibrium melting temperature, $T_0 = 505$ K. So in this case the following pressure in the crystallite corresponds to the equilibrium of crystal and liquid phases

$$p_{SL} = p_{atm} - \frac{2\sigma_{SL}}{r_*(v_L/v_S - 1)} = -3.2 \text{ GPa} . \tag{3.15}$$

Here p_{atm} is the atmospheric pressure, σ_{SL} is the surface tension of the crystal–liquid boundary, r_* is the radius of the critical crystal nucleus, and v_L and v_S are, respectively, the specific volumes of the supercooled melt and the crystal. The estimate for the pressure, p_{SL}, of liquid–crystal phase equilibrium at planar interfaces, obtained by extrapolating the melting curve of tin by using the Simon approximation [78], results in the close value $p_{SL}\,(T = 384 \text{ K}) = -3.46$ GPa. The observed kinetics of crystal nuclei formation in supercooled melts [6] corresponds to the theoretical predictions and confirms the assumption that the melting curve can be extended far beyond the triple point.

For "normally" melting substances, the derivative dp/dT_{SL} along the melting curve is positive. In this case, the analytic extension of the melting curve beyond the triple point leads to a limiting pressure of the equilibrium curve equal to

$$p\,(T \to 0) = -p_* . \tag{3.16}$$

The value $p_* > 0$ can be considered as a specific characteristic parameter characterizing the substance with respect to the melting behavior. According to the requirements of Nernst's theorem, the low-temperature course of the melting curve should satisfy the condition

$$\left(\frac{dp}{dT_{SL}}\right) \to 0 \qquad \text{at} \qquad T \to 0 . \tag{3.17}$$

At $T \to 0$, from Eqs. (3.14) and (3.16) the result

$$p_* = \left(\frac{\Delta u_{SL}}{\Delta v_{SL}}\right)_{T \to 0} \tag{3.18}$$

is obtained, where $\Delta u_{SL} = u_L - u_S$ and $\Delta v_{SL} = v_L - v_S$, i.e., the value p_* has the meaning of a characteristic internal pressure in the low-temperature limit (see Section 3.5). This value is close to the theoretical limit of stability of the body at uniform tension and $T \to 0$ [99].

In contrast to liquid–vapor phase equilibrium, the melting of simple substances can be investigated up to very high pressures. The absence of a universal feature like the critical point on the melting curve complicates the description of the phase equilibrium from the point of view of thermodynamic similarity. The main obstacle is connected with the uncertainty in the choice of the scaling parameters for the thermodynamic variables. The pressure in the triple point, for example, cannot be used as a scaling parameter for the melting curve. As already pointed out, the absence of a critical point is due to the essential nonremovable discontinuity in the properties of crystal and liquid. The properties of each of the considered condensed phases are described by its own equation of state. This fact is a reflection of the existence of two alternative ways of dense packing of atoms, the regular and irregular ones, having a different symmetry. The consideration of the low-temperature asymptotics of the melting curve, performed in agreement with the behavior of the crystal–liquid system in a wide range of temperatures at positive pressures, $p > 0$, gives a key to the generalization of the data on melting from the point of view of thermodynamic similarity. The value p_* can be used as a characteristic scaling parameter in the description of the melting process.

Formally, the Simon equation Eq. (3.3) obeys the low-temperature asymptotics, satisfying the conditions Eq. (3.16) and (3.17). The value $-p_*$ in Eq. (3.3) has the meaning of the pressure on the melting curve at $T = 0$. The parameters of this equation, p_* and c, are determined by processing the experimental data. In this case the construction of the metastable extension of the melting curve using Eq. (3.3) turns out to be in agreement with the behavior of the crystal–liquid system in a wide temperature range of pressures $p > 0$. The form of Eq. (3.3) gives an indication that the choice of characteristic temperature T_0 and characteristic pressure p_* is reasonable. By introducing a shifted pressure scale via

$$p' = p + p_* \,, \tag{3.19}$$

one can write Eq. (3.3) in the universal form [100]

$$\frac{p'_1}{p'_2} = \left(\frac{T_1}{T_2}\right)^c \,, \tag{3.20}$$

where the indices 1 and 2 are related to arbitrary points on the melting curve. Going over to reduced values for temperature and pressure defined via

$$\widetilde{p}_{SL} = \frac{p'}{p_*} \,, \qquad \widetilde{T}_{SL} = \frac{T}{T_0} \,, \tag{3.21}$$

we obtain the Simon equation in the following form

$$\widetilde{p}_{SL} = \widetilde{T}^c_{SL} \,. \tag{3.22}$$

This equation is a particular realization of a one-parameter equation of the type of Eq. (2.9) in application to melting curves. The temperature T_0 corresponds to an equilibrium coexistence of crystal and liquid at $p = 0$ or $p' = p_*$.

The procedure of introducing the shifted scale for the pressure (Eq. (3.19)) is similar to the transition from temperature, t, in the Celsius scale to the temperature, T, in the perfect gas scale (which coincides with the absolute temperature scale) via $T = T_0 + t$. The expression for the pressure of a perfect gas at constant volume

$$p = p_0\,(1 + \alpha t) \quad \text{with} \quad \alpha = 1/273 = T_0^{-1} = 0.00366\ \text{K}^{-1} \tag{3.23}$$

then takes the form $p/p_0 = T/T_0$, similar to Eq. (3.22) (where $p_* = 0$, $c = 1$). The difference is that the temperature scale of a perfect gas is universal, while the new pressure scale p' is individual for each substance.

The application of the scaling values of temperature (T_0) and pressure (p_*) and reduced values of these variables $\widetilde{p}_{SL}$ and $\widetilde{T}_{SL}$ allows us to speak about the thermodynamic similarity of melting. The point with the coordinates $\widetilde{p}_{SL} = 1$ and $\widetilde{T}_{SL} = 1$ serves as a reference point for the melting curves of various substances. The role of a similarity parameter is played by the exponent c in Eqs. (3.20) and (3.22). Substances with equal (close enough) values of the parameter c form a group of similar substances with respect to their melting behavior. It can be easily shown that the dimensionless angular coefficient of the melting curve as given by Eq. (3.3) is constant along the melting curve and equal to the parameter c, i.e., the relation

$$\frac{T}{p'}\frac{dp}{dT_{SL}} = c \tag{3.24}$$

holds.

In addition to the Simon equation, other realizations of one-parameter correlations of the type of Eq. (2.9) for the melting curve can be proposed [101], for example

$$\widetilde{p}_{SL} = \widetilde{T}^{a}_{SL}\exp[b(\widetilde{T}_{SL} - 1)]\ , \tag{3.25}$$

$$\widetilde{p}_{SL} = \widetilde{T}^{a}_{SL}(1 + b_1\widetilde{T}_{SL} + b_2\widetilde{T}^{2}_{SL})\ . \tag{3.26}$$

Both equations satisfy the low-temperature asymptotic relations Eqs. (3.16) and (3.17). Taking into account the relations $\widetilde{p}_{SL} = 1$ at $\widetilde{T}_{SL} = 1$, in Eq. (3.26) the identity $b_1 = -b_2 = b$ holds. According to the definition Eq. (3.24), we have $c = a + b$ for Eq. (3.25) and $c = a - b$ for Eq. (3.26). The verification of Eqs. (3.25) and (3.26) has shown the acceptability of these approximations for the melting curves, but does not reveal any noticeable advantages as compared to the much simpler equation Eq. (3.22).

One can establish a correlation of the similarity parameter c with the caloric parameters Δs_{SL}, Δh_{SL}, and Δu_{SL} characterizing melting (cf. Eq. (3.14)). Taking into account the Clausius–Clapeyron equation Eq. (1.2), Eq. (3.24) for the point $p = 0$ and $T = T_0$ can be rewritten in the form

$$c = \frac{T_0}{p_*}\left(\frac{\Delta s_{SL}}{\Delta v_{SL}}\right)_{p=0} = \frac{1}{p_*}\left(\frac{\Delta h_{SL}}{\Delta v_{SL}}\right)_{p=0}\ . \tag{3.27}$$

For $p = 0$, we have $\Delta h_{SL} = \Delta u_{SL}$, and for $T = 0$ the result $\Delta h_{SL} = 0$ follows. Consequently, we have

$$\Delta u_{SL}\,(T = 0) = -p\Delta v_{SL}\,(T = 0) = p_*\Delta v_{SL}\,(T = 0)\ . \tag{3.28}$$

For the similarity parameter c we obtain then from Eq. (3.27)

$$c = \left(\frac{\Delta u_{SL}}{\Delta v_{SL}}\right)_{p=0} \Big/ \left(\frac{\Delta u_{SL}}{\Delta v_{SL}}\right)_{T=0}\ . \tag{3.29}$$

If we introduce the dimensionless jumps of entropy and volume of melting via

$$\Delta\widetilde{s}_{SL} = \frac{\Delta s_{SL}}{R} , \qquad \Delta\widetilde{v}_{SL} = \frac{\Delta v_{SL} p'}{RT} , \tag{3.30}$$

then according to the Clausius–Clapeyron equation Eq. (1.2) we get from Eq. (3.24) another relation for the similarity parameter of the form

$$c = \frac{\Delta\widetilde{s}_{SL}}{\Delta\widetilde{v}_{SL}} . \tag{3.31}$$

Equations (3.27), (3.29) and (3.31) allow one a better understanding of the meaning of the parameter c. It depends both on the forces of attraction and forces of repulsion between the atoms or molecules of the substance under consideration as well as on configurational properties of the coexisting phases.

3.5 Internal Pressure in a Liquid Along the Equilibrium Curves with Crystal and Vapor

The internal pressure, p_i, of an isotropic phase is determined by the derivative of the internal energy, u, with respect to volume as

$$p_i = \left(\frac{\partial u}{\partial v}\right)_T . \tag{3.32}$$

Employing the van der Waals equation of state, we have as a special case

$$p_i = 3p_c \left(\frac{v_c}{v}\right)^2 . \tag{3.33}$$

In a system in thermodynamic equilibrium, the dependence between internal pressure, p_i, and external pressure, p, is given by the relation

$$p_i = T\left(\frac{\partial p}{\partial T}\right)_v - p , \tag{3.34}$$

where

$$p_t = T\left(\frac{\partial p}{\partial T}\right)_v \tag{3.35}$$

is called the thermal or full pressure. The behavior of the internal pressure in the course of changes of the state of the system reflects the change in the balance of the forces of attraction (resulting in $p_i > 0$) and repulsion ($p_i < 0$) in taking the average with respect to the positions of all particles. The pressures p_t and p_i in various states of the system can be calculated by using Eqs. (3.34) and (3.35), if the thermal equation of state of the substance is known.

Employing the Clausius–Clapeyron equation Eq. (1.2) and the relation

$$T\Delta s = \Delta h = \Delta u + p\Delta v , \tag{3.36}$$

where h is the enthalpy, another parameter $\hat{p}$ can be introduced, which has the dimension of pressure and characterizes the crystal–liquid phase transition. This parameter is defined as

$$\hat{p}_{SL} \equiv \left(\frac{\Delta u_{SL}}{\Delta v_{SL}} \right) = T \frac{dp}{dT_{SL}} - p \,. \tag{3.37}$$

The corresponding notation for liquid–vapor phase transitions can be introduced similarly.

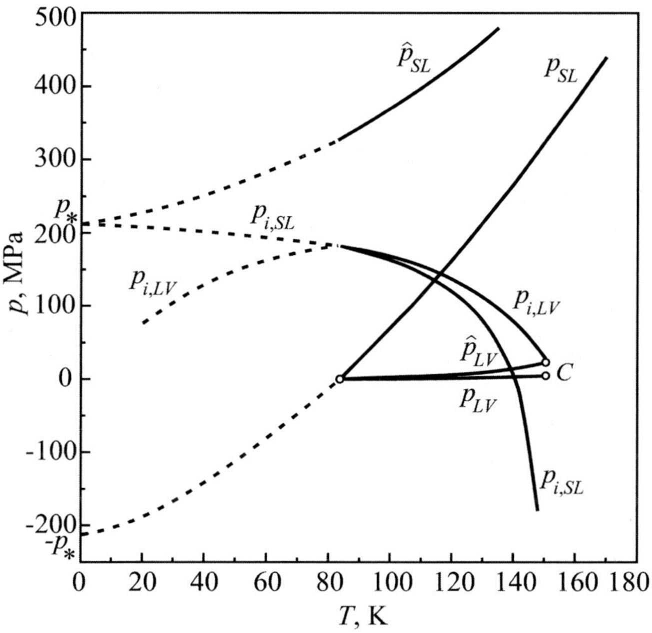

Figure 3.5: The behavior of the internal pressure, p_i, in a liquid and the quantity $\hat{p}$ along the curves of crystal–liquid phase equilibrium, $p_{SL}(T)$, and liquid–vapor, $p_{LV}(T)$. By (C) the critical point of the liquid–vapor coexistence curve is denoted. The dashed parts of the curves represent the continuation of the coexistence curves into the region of metastable states beyond the triple point. The results shown here are for argon.

Equations(3.34) and (3.37) have a similar form. However, the specific volumes of a liquid, v_L, on the curves of crystal–liquid and liquid–vapor coexistence change differently with the increase of temperature: in the first case $dv_L/dT_{SL} < 0$ holds, while in the second case the opposite inequality $dv_L/dT_{LV} > 0$ is fulfilled. There are also differences in the relative slope of the phase coexistence curve on the (T, p)-plane and of the family of isochores in the points where the isochore is attached to these curves. For the crystal–liquid coexistence curves, we have

$$\frac{dp}{dT_{SL}} > \left(\frac{\partial p}{\partial T} \right)_v \,, \tag{3.38}$$

while for the liquid–vapor coexistence curve

$$\frac{dp}{dT_{LV}} < \left(\frac{\partial p}{\partial T}\right)_v \tag{3.39}$$

holds. These results imply that, if we take into account Eqs. (3.34) and (3.37), the value $\hat{p}_{SL}$ is larger and $\hat{p}_{LV}$ is smaller than the corresponding internal pressures, p_i, in the liquid along the phase equilibrium curves. Taking into consideration the low-temperature region of metastable states of coexisting phases, we may note that the relations

$$\hat{p}_{LV} \to 0\,, \qquad p_{i,LV} \to 0 \qquad \text{at} \qquad T \to 0\,, \tag{3.40}$$

$$\hat{p}_{SL} \to p_*\,, \qquad p_{i,SL} \to p_* \qquad \text{at} \qquad T \to 0 \tag{3.41}$$

hold both in the same limit. These results are a consequence of Eqs. (3.34), (3.37), and (3.3). At any arbitrary point of the melting curve, we have

$$\hat{p}_{SL} = cp_* + (c-1)\,p\,. \tag{3.42}$$

The curves $p_{i,LV}(T)$ and $p_{i,SL}(T)$, referring to the liquid, intersect in the triple point.

Figures 3.5 (for argon) and 3.6 (for sodium) show the behavior of the functions $p_{SL}(T)$, $p_{LV}(T)$, $p_i(T)$, and $\hat{p}(T)$ of the liquid phase along the liquid–crystal and liquid–vapor equilibrium curves. The curves $p_i(T)$ and $\hat{p}(T)$ were plotted by using (T, p, v)-data from Ref. [102] for argon and from Refs. [103, 104] for sodium. The melting curves were extended into the region of negative pressures, $p < 0$, according to Eq. (3.3). From Figs. 3.5 and 3.6 it can be seen that the internal pressure in a liquid $p_{i,SL}$ coincides at $T \to 0$ with the pole, p_*, and compensates the stress of uniform tension $p = -p_*$ when approaching the limiting low-temperature point of the melting curve. This corresponds to the theoretical solidity of the condensed phase at $T \to 0$ and implies the proximity of the coexisting phases to the spinodal states. Note that the value of the internal pressure of the crystal phase on the melting curve is close to that of the liquid. The temperature dependencies of p_i for both phases along the melting curve are almost identical.

With the increase of temperature and pressure, the quantities $\hat{p}_{SL}$ and $p_{i,SL}$ diverge rapidly. This different type of behavior is caused by the absence of some point (like the critical point in liquid–vapor coexistence) along the crystal–liquid coexistence curve, where the equilibrium coexistence is terminated. The existence of the critical point for liquid–vapor equilibrium leads to a conversion of the curves $p_{i,LV}$ and $\hat{p}_{LV}$ with an increase of temperature and to their coincidence at the critical point. Along the whole liquid–vapor equilibrium curve, the internal pressure is positive, $p_i > \hat{p}_{LV} > 0$, while on the melting curve the internal pressure passes through zero with the increase of temperature and becomes negative. This kind of behavior gives a proof of the well-known point of view that the liquid–crystal phase transition is not connected with a dominant role of the forces of attraction in the system like it is in the case of gas condensation.

The previous consideration of the correlation between the quantity p_* and the internal pressure, $p_{i,L}$, in the liquid phase allows us to propose a way to estimate the pressure pole, p_*, in the Simon equation employing properties of the liquid near the melting point at atmospheric (or zero) pressure. Employing the concept of corresponding states along the melting curves,

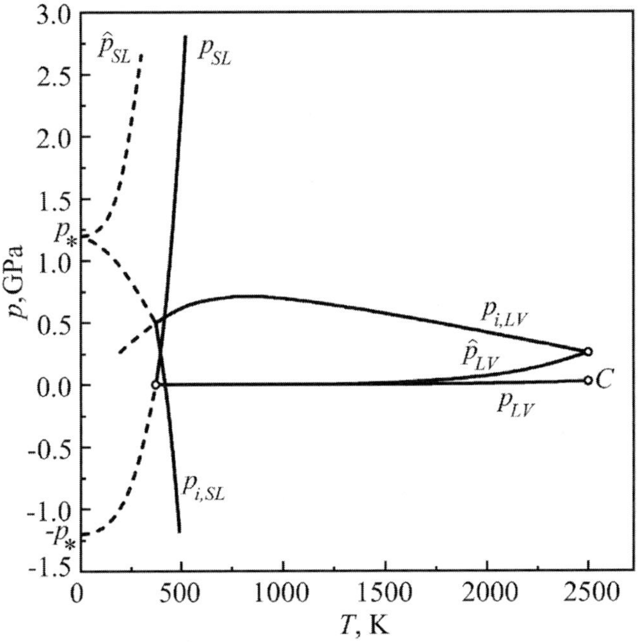

Figure 3.6: The behavior of the internal pressure, p_i, in a liquid and the quantity $\hat{p}$ along the curves of crystal–liquid phase equilibrium, $p_{SL}(T)$, and liquid–vapor, $p_{LV}(T)$. By (C) the critical point of the liquid–vapor coexistence curve is denoted. The dashed parts of the curves represent the continuation of the coexistence curves into the region of metastable states beyond the triple point. The results shown here are for sodium.

Table 3.2: Comparison of the pressures p_* and p_i (1) for a group of substances of different nature.

Substance	T_0, K	$p_{i,L}(1)$, MPa	p_*, MPa	α
Argon	83.8	190 [37]	211.4 [78]	0.90
Krypton	115.8	216 [37]	237.6 [78]	0.91
Xenon	161.4	230 [37]	261.0 [78]	0.88
Nitrogen	63.1	136 [39]	160.6 [105]	0.85
Carbon dioxide	216.5	351 [106]	355 [65]	0.99
Ammonia	195.3	521 [107]	527 [78]	0.99
Butane	134.9	340 [108]	363 [109]	0.94
Decane	243.2	305 [110]	319 [111]	0.96
Benzene	278.6	360 [112]	360 [113]	1.0
Carbon tetrachloride	250.6	318 [114]	291.9 [78]	1.1

we determine the internal pressure, $p_{i,L}$, in the liquid phase for various substances at the points of the melting curves with the same values of the reduced variables, $T/T_0 = 1$ and $p'/p_* = 1$.

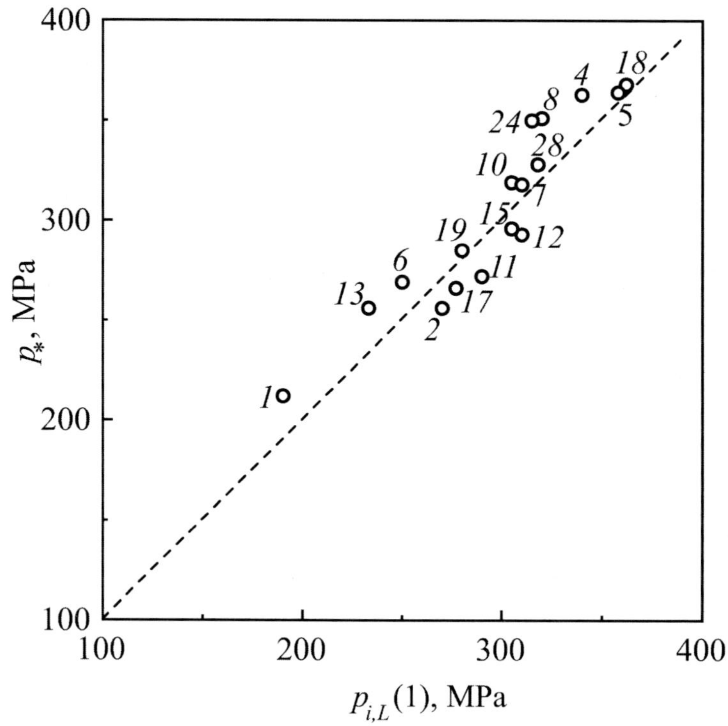

Figure 3.7: Correlation between the pressures, $p_{i,L}(1)$ and p_*, for a group of saturated hydrocarbons. The numbers refer to the number of carbon atoms in a molecule of the substance. The dashed curve corresponds to the curve $p_* = p_{i,L}(1)$ (see also Table 3.3).

The values of $p_{i,L}(1)$ are computed with Eq. (3.34) using experimental (T, p, v)-data. For thermodynamically similar substances in corresponding states, we propose the ratio

$$\frac{p_{i,L}(1)}{p_i(0)} = \frac{p_{i,L}(1)}{p_*} = \alpha \tag{3.43}$$

to be the same. This assumption can be verified for well-studied substances. Employing further this result, the value of the parameter p_* can then be estimated by the following relation

$$p_* = \frac{p_{i,L}(1)}{\alpha} \, . \tag{3.44}$$

Table 3.2 gives the values of the parameter α, the pressures p_* and $p_{i,L}(1)$ for a number of substances of very different nature. It also contains the references to the publications from where the (T, p, v)-data for the calculation of $p_{i,L}(1)$ and the data for the melting curves, employed for the determination of p_*, were taken. It follows from the table that the values of α for monatomic and diatomic substances are close to 0.9. For substances with more complex molecular structure, the values of the pressure parameters p_* and $p_{i,L}(1)$ become closer. The

Table 3.3: Thermodynamic properties characterizing melting of a group of saturated hydrocarbons at atmospheric pressure, and the parameters of their melting curves.

	T_0, K	Δh_{SL}, kJ/kg	Δv_{SL}, 10^{-3}m^3/kg	$p_{i,L}\,(1)$, MPa	p_*, MPa	c_{exp}	c_{calc}
C_1	90.7	58.7 [105]	0.166 [105]	190 [108]	212 [105]	1.69	1.86
C_2	90.4	95.3 [59]	0.172 [59]	270 [108]	256 [115]	2.18	2.05
C_4	134.9	80.3 [116]	0.100 [116]	340 [108]	363 [109]	2.21	2.36
C_5	143.2	116.6 [111]	0.131 [111]	360 [117]	366 [111]	2.41	2.47
C_6	178.6	152.1 [111]	0.171 [111]	250 [118]	269 [111]	3.12	3.19
C_7	182.7	140.6 [111]	0.13 [111]	310 [119]	318 [111]	3.18	3.49
C_8	216.2	181.9 [111]	0.176 [111]	320 [120]	351 [111]	2.81	2.94
C_{10}	243.2	202.0 [111]	0.170 [111]	305 [110]	319 [111]	3.48	3.90
C_{11}	247.7	142.2 [121]	0.123 [121]	290 [122]	272 [121]	4.16	3.99
C_{12}	263.7	195.7 [122]	0.196 [122]	310 [122]	293 [121]	3.57	3.43
C_{13}	267.7	154.7 [121]	0.144 [121]	233 [123]	256 [121]	4.45	4.62
C_{15}	282.7	163.3 [121]	0.132 [121]	300 [124]	296 [121]	4.03	4.05
C_{17}	294.7	168.4 [121]	0.136 [121]	277 [124]	266 [121]	4.43	4.44
C_{18}	301.2	242.1 [125]	0.208 [121]	362 [126]	368 [121]	3.29	3.22
C_{19}	304.7	170.7 [121]	0.140 [121]	280 [127]	284 [121]	4.36	4.36
C_{24}	323.2	161.5 [125]	0.127 [121]	317 [128]	350 [121]	3.69	4.01
C_{28}	334.4	163.7 [125]	0.133 [121]	318 [129]	328 [130]	3.75	3.87

value of α for them is close to one. This result implies that the internal pressure in the liquid, $p_{i,SL}$, varies only slightly along the melting curve in a temperature range from T_0 to 0 K. The slight variation of the internal pressure, p_i, when the temperature increases from zero to $T \geq T_0$ is due to the partial compensation of the two terms in Eq. (3.34).

Figure 3.7 gives an illustration of the degree of accuracy of the dependence, Eq. (3.44) between the pressure parameters p_* and $p_{i,L}\,(1)$ for a large group of saturated hydrocarbons. Taking into account the inaccuracy in the determination of these parameters, the value of α can be considered as equal to one to a very good approximation. Table 3.3 contains the values of the pressures p_* and $p_{i,L}\,(1)$. Here the values of volume, Δv_{SL}, and enthalpy jumps, $\Delta h_{SL} = T\Delta s_{SL}$, in melting at atmospheric pressure are also given.

The correlation, illustrated in Fig. 3.7, allows us to estimate the value of the pressure pole, p_*, in the Simon equation, if the (T, p, v)-properties of the liquid phase near the melting temperature T_0 are known. The parameter c in the Simon equation can be determined by the data for Δv_{SL} and Δh_{SL}, using Eq. (3.27), via

$$c = \frac{\alpha \Delta h_{SL}}{p_{i,L}\Delta v_{SL}} \,. \tag{3.45}$$

Table 3.3 also contains the values of the parameters of the Simon equation calculated by using Eq. (3.45) (denoted as c_{calc} in Table 3.3) and experimental values of this parameter c_{exp} determined directly from the data for the melting curves.

Table 3.4: Comparison of the values of the internal pressure, p_i, in the crystalline and liquid phases at the melting point at atmospheric pressure for simple substances.

Substance	T_0, K	$p_{i,L}(1)$, MPa	$p_{i,S}(1)$, MPa
Neon	24.55	70 [37]	88 [37]
Argon	83.8	190 [37]	220 [37]
Krypton	115.7	216 [37]	254 [37]
Xenon	161.3	230 [37]	256 [37]
Sodium	370.8	520 [103]	495 [103]
Potassium	335.7	237 [131, 132]	265 [131, 133]
Rubidium	311.9	218 [131, 134]	224 [131, 133]
Cesium	302.9	148 [131, 135]	138 [131, 133]
Mercury	234.3	1100 [136]	1150 [136]

The tangents of the isochores, $(\partial p/\partial T)_v$, for the crystalline and liquid phases to any definite point along the melting line turn out to be near to each other with respect to their values near the triple point. This similarity leads to similar values of the internal pressure in the coexisting phases. In Table 3.4, estimates of the internal pressures $p_{i,L}(1)$ and $p_{i,S}(1)$ are given for simple substances including some metals at $T/T_0 = 1$ and $p'/p_* = 1$, respectively. These data have been computed based on the experimental results for the coefficient of thermal expansion, $\alpha_p = (1/v)(\partial v/\partial T)_p$, and the isothermal compressibility, $\beta_T = -(1/v)(\partial v/\partial p)_T$, employing the relation

$$p_i = T\frac{\alpha_p}{\beta_T} - p \; . \tag{3.46}$$

Equation (3.46) is a consequence from Eq. (3.34) if one utilizes, in addition, the thermodynamic identity

$$\left(\frac{\partial v}{\partial p}\right)_T \left(\frac{\partial p}{\partial T}\right)_v \left(\frac{\partial T}{\partial v}\right)_p = -1 \; . \tag{3.47}$$

In Table 3.4, the references for the data for the thermal expansion coefficient and the isothermal compressibility are given. For the crystalline phases of potassium, rubidium, and cesium the isothermal compressibility was computed employing data for the elasticity coefficients obtained from measurements of the speed of sound [133]. Within the experimental errors, the values for the parameters $p_{i,L}$ and $p_{i,S}$ coincide. However, for noble gases, which are investigated with higher precision, the expected inequality $p_{i,L}(1) < p_{i,S}(1)$ is reconfirmed. This result is consistent with the idea about a higher elasticity of the crystalline phase as compared with the liquid and the validity of the inequality $(\partial u/\partial v)_{T,S} > (\partial u/\partial v)_{T,L}$.

Having at one's disposal the data for the thermal expansion coefficient and the isothermal compressibility of the crystalline phase of the metals, we may estimate the internal pressure $p_{i,S}(1) \cong p_{i,L}(1)$ and compare these values with the pressure pole, p_*, of the respective substance. In Table 3.5, data for the internal pressure, $p_{i,S}(1)$, at the melting point for a

Table 3.5: Comparison of the pressures p_* and $p_{i,S}(T_0)$ for different metals.

Substance	T_0, K	p_*, MPa	$p_{i,S}(1)$, MPa	$p_{i,S}(1)/p_*$
Sodium	370.8	1197 [78]	520	0.43
Potassium	335.7	427 [78]	265	0.62
Rubidium	311.9	395 [78]	224	0.57
Cesium	302.9	267 [78]	138	0.56
Aluminum	933	7000 [78]	5550	0.79
Copper	1358	13400 [137]	10000	0.75
Lead	600.2	3230 [78]	2300	0.71
Silver	1235	9200 [137]	8000	0.87
Gold	1338	10800 [137]	9885	0.92
Cadmium	594.1	4500 [78]	3000	0.67
Nickel	1728	16570 [72]	15500	0.94
Zinc	692.7	6000 [78]	3900	0.65

group of metals are given computed employing data for α_p from Ref. [131] and data for β_T obtained from the elastic properties of the crystals given in Ref. [133]. In the table, the values of the parameter p_* for the melting curve of the metals are also given. A comparison of the pressures $p_{i,S}(1)$ and p_* gives an indication of the existence of a correlation between them. This correlation between $p_{i,S}(1)$ and p_* is shown in Fig. 3.8. For alkali metals, having a bcc crystal structure, the ratio $\alpha = p_{i,S}/p_*$ is of the order of 0.5–0.6. For the other metals, having dense crystal structures, the parameter α varies in the range 0.7–0.9. This ratio is near to the ratio $p_{i,L}(1)/p_*$ for noble and two-atomic gases.

From the analysis performed here it can be concluded that the parameter p_* in the Simon equation has values close to the internal pressure in the liquid near the triple point. Note that the existence of a relationship between p_* and the internal pressures of the liquid and crystal phases along the phase equilibrium curve has been mentioned already by Simon himself [138].

3.6 Stability of Thermodynamic States and the Metastable Continuation of the Melting Curves

When the melting curves are extended beyond the crystal–liquid–gas triple point, (T_{tr}, p_{tr}), the question concerning the thermodynamic stability of the coexisting phases arises. In the range $0 < T < T_{tr}$ and $-p_* < p < p_{tr}$ or $0 < p' < p'_{tr}$, crystal and liquid are metastable with respect to the vapor. Applying a uniform tensile stress, $\sigma = -p > 0$, to the condensed system we can extend the melting curve into the region of negative pressures without violating the conditions of liquid–crystal phase equilibrium. At negative pressures, $p < 0$, a vapor phase cannot exist but the problem of metastability remains with respect to the homogeneity of crystal and liquid, i.e., with respect to the formation of cavitation bubbles. The known

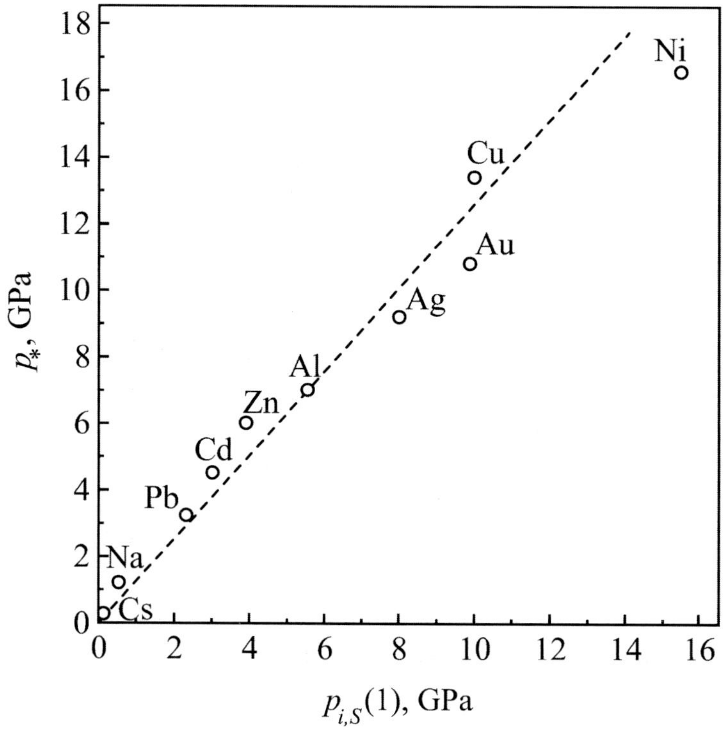

Figure 3.8: Correlation between the pressures $p_{i,S}(1)$ and p_* for metals.

estimates of the tensile strength of crystalline and liquid phases and the limits of their stability allow us to advance far into the region of negative pressures.

The boundary of thermodynamic stability with respect to small fluctuations of the homogeneous phase is determined by zero values of the elasticity coefficient, $-(\partial p/\partial v)_T$, or of the inverse coefficient of isothermal compressibility, $\beta_T^{-1} = -v(\partial p/\partial v)_T$. The condition

$$\left(\frac{\partial p}{\partial v}\right)_T = 0 \tag{3.48}$$

determines the mechanical spinodal curve. The analysis of the behavior of the elasticity coefficients of the liquid and crystalline phases along the coexistence curve allows us to conclude that the thermodynamic stability of liquid and crystal on the melting curve decreases with the decrease of temperature [100, 139–141]. The spinodal of the liquid originates from the critical point (T_c, p_c, v_c) and is located for $T \leq 0.9T_c$ in the region of negative pressures [1]. According to the van der Waals equation of state, the spinodal curve approaches the pressure $p_{sp}(0) = -27p_c$ at $T \to 0$.

The (p, T)- and (p, v)-phase diagrams for argon are shown in Figs. 3.9 and 3.10. They allow us to come to more definite conclusions concerning the relative location and behavior of the low-temperature segments of the melting curve and the liquid spinodal. The continuation

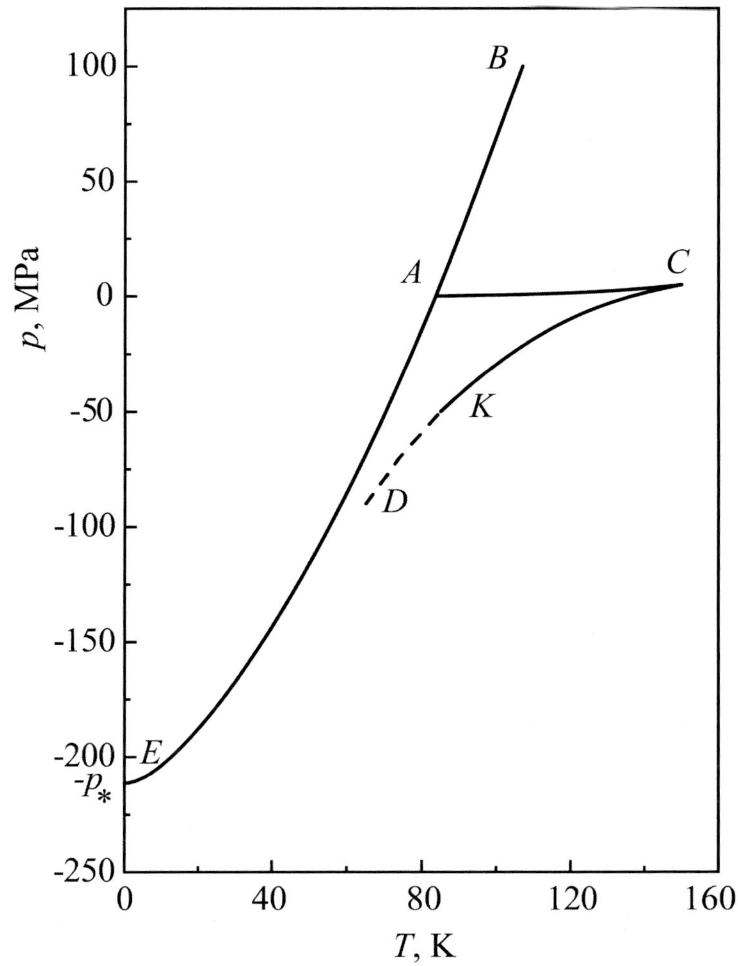

Figure 3.9: The melting curve of argon (AB) with its metastable continuation (AE) into the range of negative pressures. (CKD) is the spinodal of the liquid under tensile stress and (AC) is the liquid–vapor coexistence curve.

(AE) of the melting curve (BA) beyond the triple point corresponds to the Simon equation Eq. (3.3) with parameter values $p_* = 211.4$ MPa, $c = 1.593$, and $T_0 = 83.8$K [78]. The liquid spinodal (CK) is plotted using experimental (T, p, v)-data [142] for the range of stable and metastable states of liquid argon. The extension of the spinodal curve (KD) is less reliable. From Figs. 3.9 and 3.10 one can draw the qualitative conclusion that the melting curve converges with the liquid spinodal with the increase of the tensile stress applied to the coexisting liquid and crystalline phases.

The stability of the crystal phase is also decreased. Figures 3.11 (for argon) and 3.12 (for sodium) show the behavior of the elasticity $-(\partial p/\partial v)_T$ of the crystalline and liquid phases

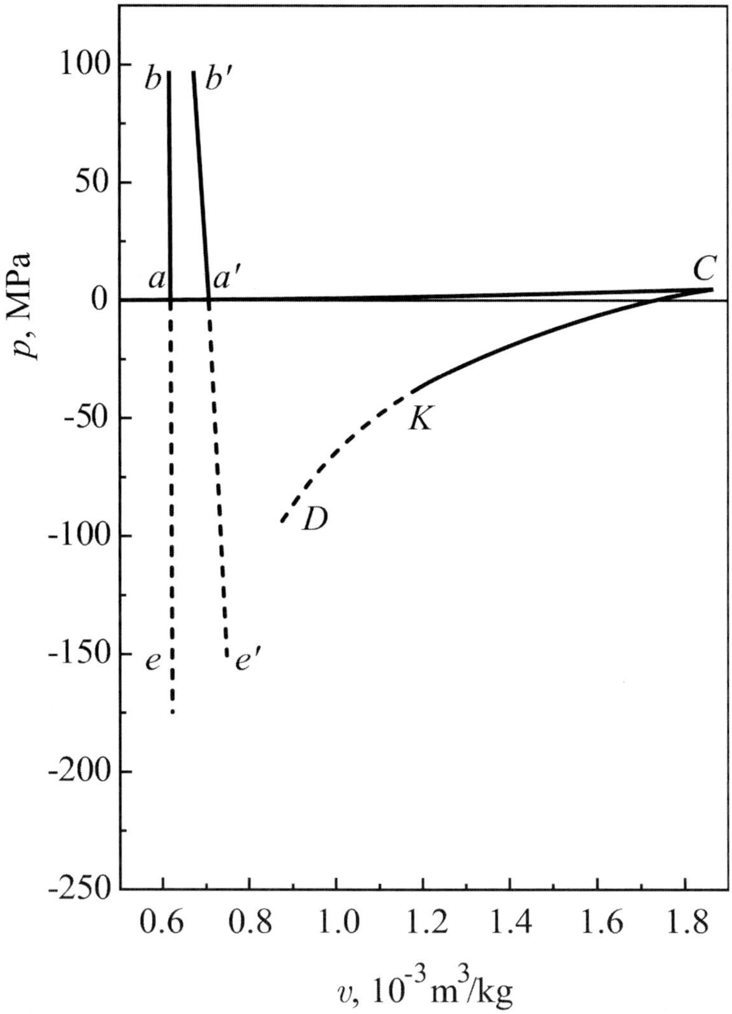

Figure 3.10: The binodal curves of crystalline (ab) and liquid (a′b′) argon with the continuations (ae) and (a′e′) into the range of negative pressures. (a′C) is the binodal curve of the liquid in equilibrium coexistence with the vapor. (CKD) is the spinodal of liquid argon under tensile stress.

along the melting curve. The liquid and crystal branches of elasticity were plotted by using data from Ref. [143] for argon and from Ref. [103] for sodium. In drawing the figures, the shifted pressure scale $p' = p + p_*$ was employed. It can be seen that the boundaries of stability of both phases $(\partial p/\partial v)_T = 0$ are reached near to the initial point $(p' = 0, T = 0)$ of the melting curve. Apparently, the liquid loses its stability earlier than the coexisting crystal. The general character of the described tendency is proven by a comparison of the values of the limiting pressure $-p_* = p(0)$ for various substances along the melting curve and the values

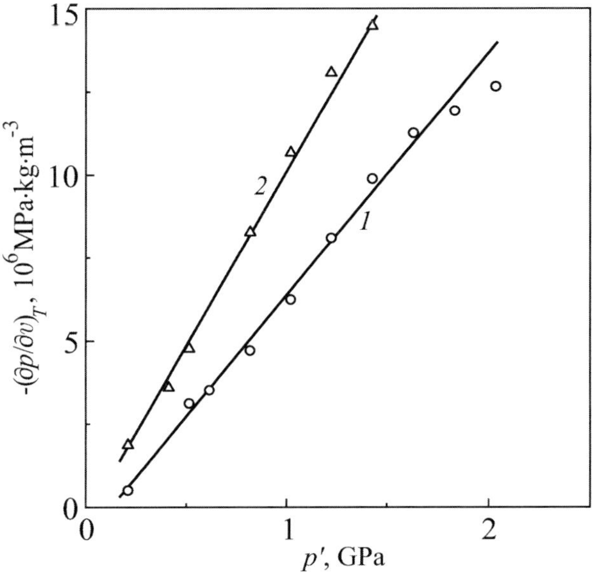

Figure 3.11: Behavior of the elasticity of liquid (1) and crystalline (2) argon on the melting curve. The pressure parameter, p_*, is $p_* = 0.21$ GPa.

of the limiting pressure $p_{sp}(0)$ on the liquid spinodal at $T \rightarrow 0$ [100, 141]. The value of $p_{sp}(0)$ can be approximated with its value, as follows from the van der Waals equation, equal to $-27p_c$.

The existence of a correlation between the asymptotic values p_* and $27p_c$ can be derived from Fig. 3.13. In this figure, on a logarithmic pressure scale, the respective data are shown for substances of different nature: inert and simple gases, hydrocarbons, fluorocarbons, associated liquids (ammonia, aniline), metals and ionic compounds (sodium chloride and potassium chloride). In the series of substances presented in the figure the values of p_* and $27p_c$ change by two or three orders of magnitude from neon and argon to copper and nickel. The numbers in the figure correspond to the number of the respective elements as given in Table 3.6. In this table, the values of p_* in the approximation of Simon's equation and $-p_{sp}(0) = 27p_c$ are collected. It also contains the parameters c and T_0 required for an application of Eq. (3.3). For most of the substances the values of p_*, c and T_0 are taken from Ref. [78] (the number of significant units is here taken to be the same as in the original reference). For some substances we calculated the parameters p_* and c employing experimental data for the melting curves. The references to the sources of the respective data are given in the far right column of the table. The values of critical pressure p_c for molecular substances and alkali metals are taken from Ref. [21]. For the other metals, estimates for p_c were used based on the results of high-temperature experiments with an impulse heating of the samples by an electric current and laser emission [144]. The estimates for p_c for sodium chloride and potassium chloride were obtained using the method of rectilinear diameter [145]. The correlation between the values

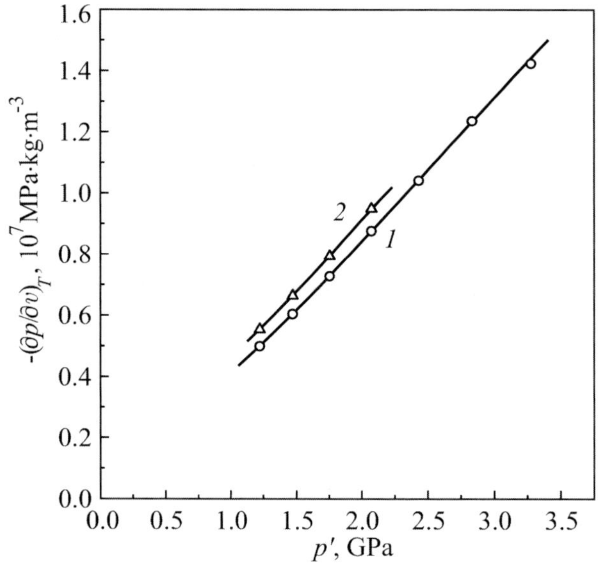

Figure 3.12: Behavior of the elasticity of liquid (1) and crystalline (2) sodium along the melting curve. The pressure parameter, p_*, is $p_* = 1.20$ GPa.

of the limiting pressures p_* and $27p_c$ is described by the following equation

$$\lg p_* = \lg (27p_c) + 0.131 \ . \tag{3.49}$$

This equation can be used for the estimation of the parameter p_* of the melting curve by the known value of the critical pressure p_c.

In order to arrive at a more complete picture as compared with Fig. 3.9, a plot of the spinodal of the crystalline phase is required. Attempts to generate the respective plots have been made by a number of authors [149–151], but the problem is still open and of great importance. To arrive at a plot of the boundaries of thermodynamic stability of liquid and crystal in a wide range of temperatures one can use (p, v, T)-data for the range of stable states. The dependence of the elasticity of a condensed phase on pressure along the isotherms is close to a linear one and, by this reason, convenient for an extrapolation into the range of metastable states. Figure 3.14 shows the dependence of the elasticity of the crystalline and liquid phases of argon on pressure at various temperatures. The curves are plotted employing (p, v, T)-data from Ref. [37]. The critical isotherm of liquid argon (curve 6) can be experimentally observed up to the spinodal state of the liquid. It tends linearly to a zero value of elasticity at the critical point ($T_c = 150.9$ K, $p_c = 4.9$ MPa). For temperatures below the critical ones, the linear extrapolation of the isotherms into the range of states with negative pressures up to zero values of the elasticity allows us to estimate the position of the boundary of thermodynamic stability. Note that in order to estimate the stability limit of a condensed phase, one can also use the almost linear dependence of the modulus of volume compressibility, $K_T = -v(\partial p/\partial v)_T$, on pressure along the isotherm. The linearity of the dependencies, illustrated in Fig. 3.14, allows

Table 3.6: Parameters of the Simon equation Eq. (3.3) and values of the pressure parameters $-p_{sp}(0) = 27p_c$ for some substances.

No.	Substance	T_0, K	p_*, MPa	c	$27p_c$, MPa	Ref.
1	Helium-4	1.11	2.06	1.547	6.0	[146]
2	Neon	24.55	103.8	1.600	71.6	[78]
3	Argon	83.81	211.4	1.593	131.3	[78]
4	Krypton	115.7	237.6	1.617	148.4	[78]
5	Xenon	161.3	261.0	1.589	157.2	[78]
6	Nitrogen	63.15	161.2	1.783	91.8	[105]
7	Oxygen	54.38	273.3	1.743	136.2	[78]
8	Fluorine	53.48	216.0	2.18	145.0	[147]
9	Methane	90.69	211.8	1.688	124.2	[105]
10	Tetrachloromethane	250.6	291.9	2.12	123.1	[78]
11	Sodium	370.78	1197.1	3.53	692.3	[78]
12	Potassium	335.7	427.0	4.44	442.7	[78]
13	Rubidium	311.9	395.1	3.74	361.7	[78]
14	Cesium	302.9	267.4	4.49	316.7	[78]
15	Aluminum	933	7000	2.20	2241	[78]
16	Nickel	1728	16570	2.53	13203	[72]
17	Copper	1357.9	13400	2.44	4563	[137]
18	Zinc	692.7	6000	2.46	6642	[78]
19	Silver	1235.2	9200	2.07	3078	[137]
20	Cadmium	594.1	4500	2.40	4050	[78]
21	Gold	1337.6	10800	2.08	3483	[137]
22	Lead	600.2	3230	2.41	4320	[78]
23	Chlorine	172.2	494	2.11	208.2	[148]
24	Iodine	386.8	467	2.74	317.3	[148]
25	Carbon dioxide	216.5	355	2.87	194.1	[65]
26	Ammonia	195.3	527	4.3	305.6	[78]
27	Ethane	90.7	256	2.18	130.1	[115]
28	Pentane	143.5	660	1.65	91.0	[111]
29	Chloroform	209.7	833	1.52	147.7	[78]
30	Bromoform	280.94	534	2.05	150.1	[78]
31	Benzene	278.6	360	2.69	132.3	[113]
32	Aniline	266.8	535	2.34	143.4	[78]
33	Nitrobenzene	278.8	620	1.93	130.0	[113]
34	Monofluorobenzene	231.0	601	1.79	122.9	[67]
35	Hexafluorobenzene	278.1	454	1.85	89.2	[67]
36	Monochlorobenzene	227.7	498	2.42	122.0	[78]
37	Monobromobenzene	242.1	480	2.46	122.0	[78]
38	Sodium chloride	1073.5	1670	2.7	945	[78]
39	Potassium chloride	1043	690	5.7	594	[78]

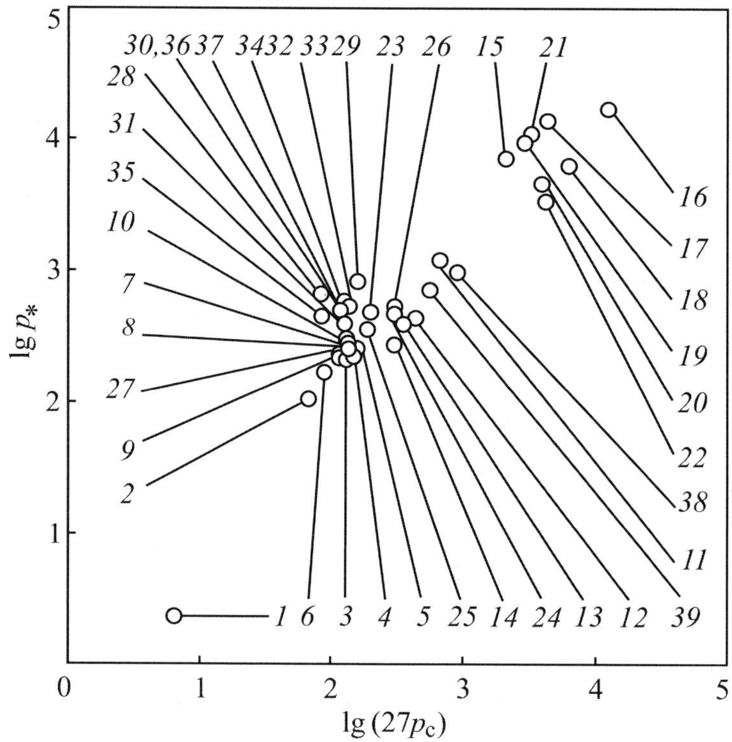

Figure 3.13: Correlation between the quantities p_* and $27p_c$ for substances of different nature. The pressure is given here in MPa. The numbers specify the same substances as in Table 3.6.

us to employ the Tait equation

$$\frac{v_0 - v}{v_0} = A \ln \left(\frac{p + B}{B} \right) . \tag{3.50}$$

The quantity $v_0 = v_0(T)$ in Eq. (3.50) corresponds to the volume at zero values of the external pressure, the parameters A and B depend also on temperature. The pressure $p = -B$ corresponds to the stability boundary Eq. (3.48), i.e., to points on the spinodal curve of the condensed phase for the given values of temperature.

Figure 3.15 shows the phase diagram of argon with the metastable extension of the melting curve into the region of negative pressures. It also shows the spinodals of crystal and liquid phases plotted using the linear extension into the region of metastable states of the dependence of the elasticity coefficient, $(\partial p/\partial v)_T$, of the condensed phase on pressure along the isotherms. The range of experimentally available (p, v, T)-data is bounded by a pressure of 100 MPa. By this reason, for the plot of the segment of the spinodal of the crystal for pressures $p > 0$ the extrapolation of the dependence of the elasticity of the crystalline phase on temperature along the isobars was employed. The nonlinearity of these curves makes the estimate of the location of the spinodal of the crystal in the range of high pressures less reliable.

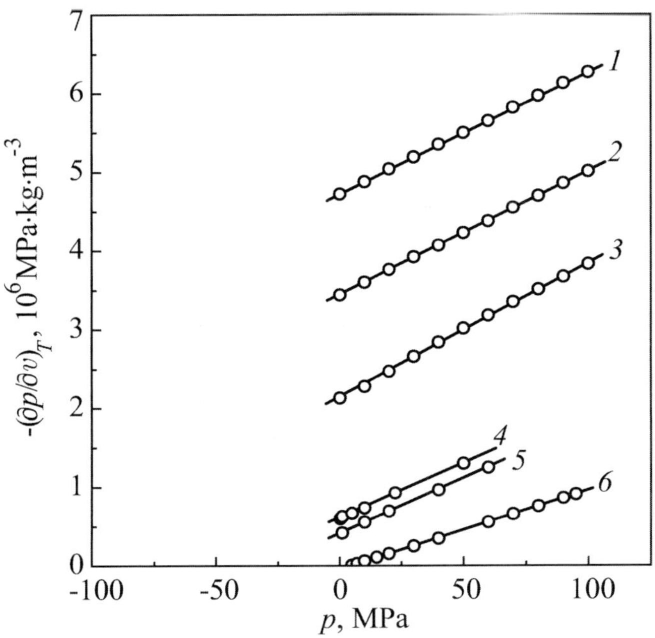

Figure 3.14: Dependence of the elasticity of crystalline (curves 1–3) and liquid (curves 4–6) argon on pressure at different temperatures: (1) 1 K, (2) 50 K, (3) 80 K, (4) 90 K, (5) 100 K, (6) 150.9 K.

The liquid isochore ($v = 0.855 \cdot 10^{-3} \mathrm{m}^3/\mathrm{kg}$), extended into the range of metastable states, coincides, within the accuracy of the method of determination of its location, with the tangent aa′ to the spinodal at the point with the coordinates $p = -30$ MPa and $T = 100$ K. This result indicates that the provided estimate of the spinodal is consistent with the condition that the spinodal is the envelope of the family of isochores in the (T, p)-plane [1].

Table 3.7: Comparison of the pressure, p_*, in the Simon equation and the constant B in the Tait equation for alkali metals at a temperature of 4.2 K.

Substance	p_*, MPa	B, MPa
Sodium	1197	1350
Potassium	427	763
Rubidium	395	489
Cesium	267	497

The relative position of the melting curve of argon in the region of negative pressures and the position of the boundaries of thermodynamic stability for the crystalline and liquid phases, shown in Fig. 3.15, give a strong indication of the possibility of the existence of a phase equilibrium of crystal and liquid in a wide range of metastable states. At low temperatures, the spinodal of the crystalline phase is located with respect to pressure significantly

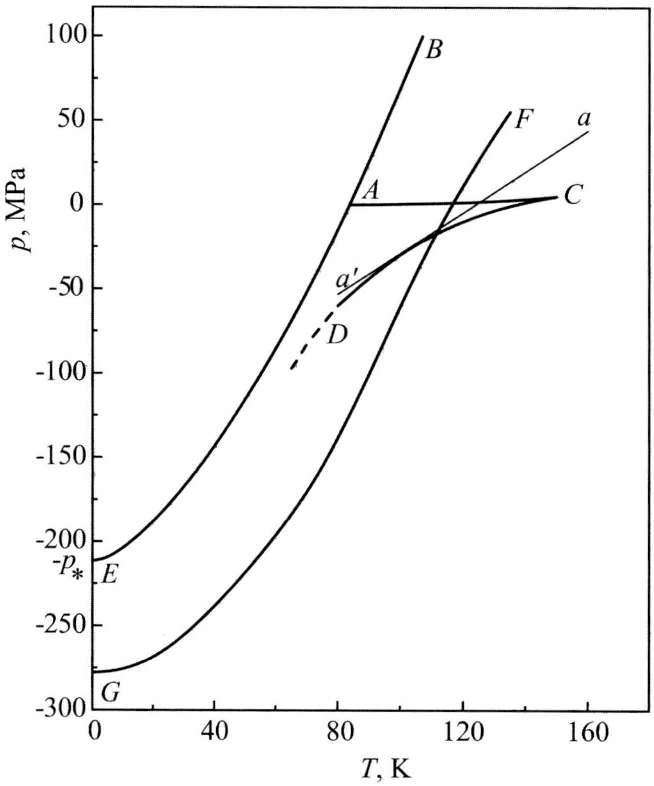

Figure 3.15: Phase diagram of argon including regions of crystal–liquid coexistence under tensile stress. Here (BAE) specifies the melting curve, (AC) the liquid–vapor equilibrium coexistence curve, (CD) the spinodal curve of the liquid, (FG) the spinodal of the crystal and (aa′) is a tangent to the spinodal curve (CD) at the point with the coordinates $p = -30$ MPa and $T = 100$ K and corresponds to the isochore of the liquid with the specific volume of $v = 0.855 \cdot 10^{-3}$ m^3/kg.

below the melting curve. It can be seen that the liquid phase on the metastable branch of the equilibrium curve also retains stability up to the states close to the low-temperature limit of the melting curve. On the phase diagram, shown in Fig. 3.15, a certain particular point can be distinguished where the condition $(\partial p/\partial v)_T = 0$ holds true for both condensed phases on the (T, p)-plane. But here the spinodal points of crystal and liquid correspond to different specific volumes of the phases.

In Fig. 3.16, the pressure dependence of the elasticity of the crystalline phases of sodium, potassium, rubidium and cesium is shown for a temperature of 4.2 K. The symbols refer to the experimental data on the compressibility of these substances in the pressure range from zero to 1 GPa according to Ref. [152]. In Table 3.7, a comparison is shown between the values of the pressure p_* and the constant B from the Tait equation Eq. (3.50) for alkali metals. This comparison shows that they are of a similar order of magnitude and gives a strong indication

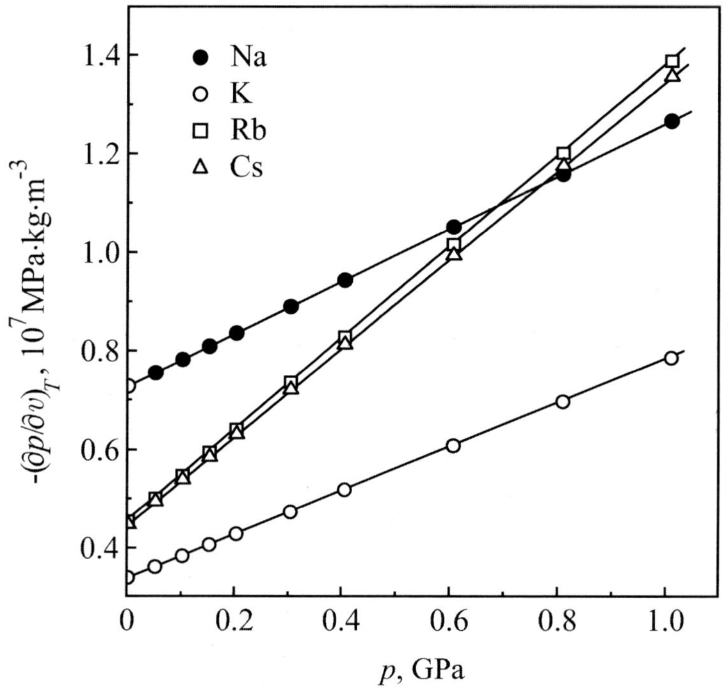

Figure 3.16: Pressure dependence of the elasticity of the crystalline phases of sodium, potassium, rubidium and cesium at a temperature of 4.2 K. The symbols refer to the experimental data, the solid lines give an interpolation according to Eq. (3.50).

of the conservation of stability of the crystalline phases of the metals under stress along the coexistence curves at $T \to 0$. One has to note, of course, that the values of the parameters B and p_* were obtained based on experimental data available for temperatures $T > 0$. In this respect, both methods of specification of these parameters are self-consistent.

Figure 3.17 shows a plot for the phase diagram of sodium. The experimental (p, v, T)-data employed for the plot [103, 153] are available only for a small temperature interval on the phase diagram. However, the segments of the spinodal curves (CD) and (FG), reestablished correctly for a wide range both for the liquid and crystalline phases, indicate that the stability of both phases is retained also in approaching the limiting point of the melting curve ($T = 0$, $p = -p_*$). Therefore, the following picture can be developed with respect to the behavior of coexisting liquid and crystal and their stability. For $p > 0$, both phases increase their stability with the increase of temperature. This conclusion is reconfirmed by an increase of the coefficient of elasticity, $-(\partial p/\partial v)_T$. No sign is found of any kind of spinodal singularity as compared with the case of liquid–vapor phase coexistence. If we extend the melting curve into the range of negative pressures of both phases, then, as can be seen from Figs. 3.11 and 3.12, the stability of the coexisting crystal and liquid phases is decreased getting closer to the stability limit given by Eq. (3.48). The range of possible coexistence of both phases under tensile stress is terminated in the low-temperature limit if the condition $(\partial p/\partial v)_T = 0$ is reached for one of the coexisting phases.

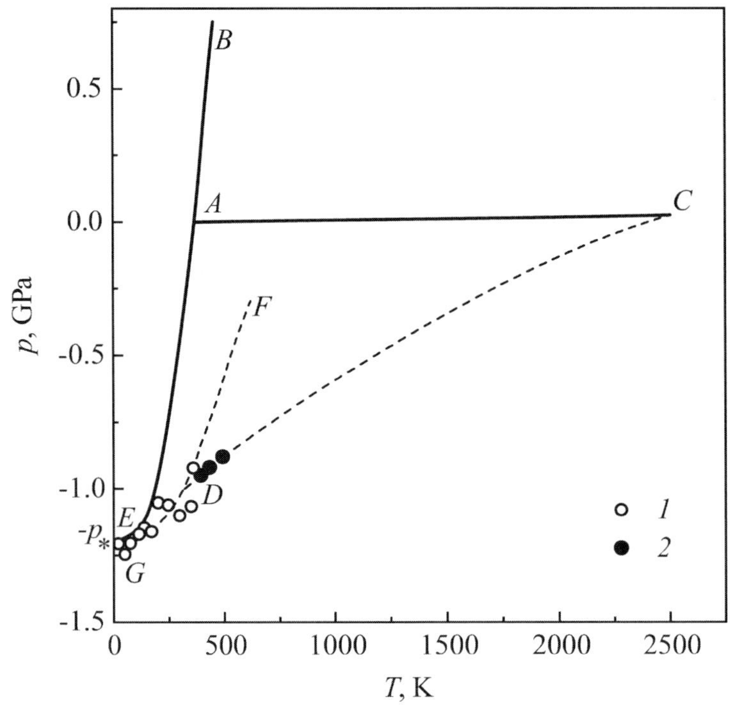

Figure 3.17: (p, T)-phase diagram of sodium including regions of crystal–liquid coexistence under tensile stress. Here (BAE) specifies the melting curve, (AC) the liquid–vapor equilibrium coexistence curve, (CD) the spinodal curve of the liquid, and (FG) the spinodal of the crystal. The circles (1) refer to the spinodal of the crystal while the dots (2) specify the spinodal of the liquid

The critical point is the upper limit of the liquid–vapor equilibrium curve. The difference in the behavior of the stability of the liquid at crystal–liquid and liquid–vapor phase transitions for argon is shown at Fig. 3.18. In this figure, the variation of the elasticity of the liquid phase is shown in dependence on temperature for the cases of both liquid–vapor (curve (LV)) and liquid–crystal (SL) phase coexistence; C is the critical point. The curve $(\partial p/\partial v)_T = 0$ determines the location of the spinodal (Sp). The deviation between spinodal and solid–liquid coexistence curve at $T = 0$ may be a result of the limited accuracy in the determination of both curves in the low-temperature limit. The behavior of the plot shown in Fig. 3.18 leads to the suggestion that the liquid spinodal connects the limiting states on the liquid–vapor and solid–liquid coexistence curves. Both mentioned states are maximally separated by temperature and are located at the boundary of stability of the coexisting phases. With respect to the stability of the liquid, a certain similarity may be noticed between the liquid–vapor critical point and the limit $T = 0$ of crystal–liquid coexistence in the region of metastable states. However, in contrast to the behavior at the critical point the mentioned limiting state for solid–liquid phase equilibrium does not imply equality of the properties of the coexisting phases and a consistent loss of stability.

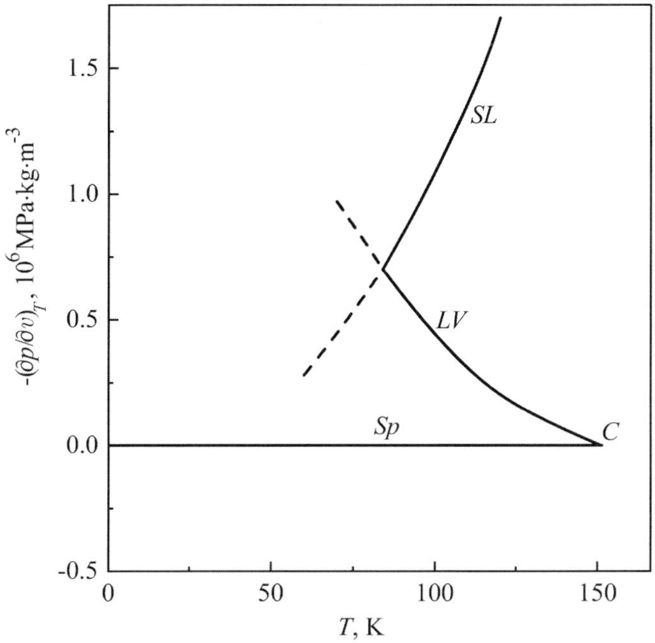

Figure 3.18: Coefficient of elasticity of liquid argon along the crystal–liquid (SL) and liquid–vapor (LV) equilibrium phase coexistence curves. (Sp) is the spinodal curve of the liquid and (C) is the liquid–vapor critical point.

The spinodal of the crystal phase can also be constructed if we have a relatively simple and physically substantiated relation for the dependence of the cohesive energy of a solid on volume. For most substances of different nature the potential energy for one gram-atom of a crystal can be presented as the sum of two power terms [154]

$$u_P = \frac{A_0}{v^{n/3}} - \frac{B_0}{v^{m/3}} , \tag{3.51}$$

with $n > m$. The coefficients A_0 and B_0 are determined by the nature of the interatomic forces and the crystal lattice structure. The first term on the right-hand side of Eq. (3.51) describes the energy of the repulsive forces, the second term describes the energy of the forces of attraction. Since, at a constant temperature, the internal pressure in a crystal is determined by the volume dependence of the potential energy we have from Eq. (3.51) (taking into account the definition Eq. (3.32)) the following notation for $p_{i,S}$

$$p_{i,S} = \frac{B}{v^{(m/3)+1}} - \frac{A}{v^{(n/3)+1}} . \tag{3.52}$$

The quantities A and B depend on temperature. They can be determined if (p, v, T)-data of the respective substance are available.

For the external pressure of a crystal, taking into account Eq. (3.52), we have

$$p = T\left(\frac{\partial p}{\partial T}\right)_v + \frac{A}{v^{(n/3)+1}} - \frac{B}{v^{(m/3)+1}} \,. \tag{3.53}$$

An analysis of experimental data shows that for most of the crystals the thermal pressure

$$T\left(\frac{\partial p}{\partial T}\right)_v = T\frac{\alpha_p}{\beta_T} \tag{3.54}$$

changes only slightly along an isotherm. The pressure and density dependence of this quantity, in a pressure range comparable with the scale p_*, can be neglected without introducing a large error. An exception from this rule are the quantum crystals composed of helium, hydrogen, or neon atoms, for which this dependence cannot be neglected.

From Eq. (3.53), taking into account the constancy of the quantity $T\left(\partial p/\partial T\right)_v$, we obtain the following relation for the isothermal compressibility

$$\frac{1}{\beta_T} = \left(\frac{n}{3}+1\right)\frac{A}{v^{(n/3)+1}} - \left(\frac{m}{3}+1\right)\frac{B}{v^{(m/3)+1}} \,. \tag{3.55}$$

From this equation, employing the condition $\beta_T^{-1} = 0$, we can derive a dependence for the crystal volume along the spinodal

$$v_{sp} = \left(\frac{n+3}{m+3}\cdot\frac{A}{B}\right)^{3/(n-m)} \,. \tag{3.56}$$

The equation of state Eq. (3.53) and Eq. (3.55), taken at zero external pressure, make it possible to determine the temperature dependences of the quantities A and B. With Eq. (3.54) we obtain

$$A = \frac{3}{(n-m)}\cdot\frac{v_0^{(n/3)+1}}{\beta_T}\cdot\left[\left(\frac{m}{3}+1\right)T\alpha_p + 1\right] \,, \tag{3.57}$$

$$B = \frac{3}{(n-m)}\cdot\frac{v_0^{(m/3)+1}}{\beta_T}\cdot\left[\left(\frac{n}{3}+1\right)T\alpha_p + 1\right] \,. \tag{3.58}$$

The values of A and B at some temperature can be calculated from the data for the volume v_0 and coefficients α_p and β_T, taken at zero (atmospheric) external pressure and this temperature. The knowledge of the quantities A and B makes it possible, employing Eq. (3.56), to calculate the volume $v_{S,sp}$ along the spinodal as

$$v_{S,sp} = \left\{\frac{n+3}{m+3}\cdot v_0^{(n-m)/3}\cdot\frac{\left(\frac{m}{3}+1\right)T\alpha_p+1}{\left(\frac{n}{3}+1\right)T\alpha_p+1}\right\}^{3/(n-m)} \,. \tag{3.59}$$

From the equation of state Eq. (3.53), utilizing Eqs. (3.56), (3.57), and (3.58), we obtain for the pressure $p_{S,sp}$ on the crystal spinodal the following relationship

$$p_{S,sp} = \frac{T\alpha_p}{\beta_T} + \frac{3}{n-m} \cdot \frac{1}{\beta_T} \cdot \frac{\left[\left(\frac{n}{3}+1\right) T\alpha_p + 1\right]^{(n+3)/(n-m)}}{\left[\left(\frac{m}{3}+1\right) T\alpha_p + 1\right]^{(m+3)/(n-m)}} \times \left[\left(\frac{m+3}{n+3}\right)^{(n+3)/(n-m)} - \left(\frac{m+3}{n+3}\right)^{(m+3)/(n-m)}\right] . \quad (3.60)$$

For the Lennard-Jones crystal with $n = 12$ and $m = 6$ at a temperature $T = 0$ K we obtain from Eqs. (3.59) and (3.60) the following relation for the volume and the pressure on the spinodal

$$v_{S,sp}(0) = 1.29v_0 , \qquad p_{S,sp}(0) = -0.093\beta_T^{-1} , \quad (3.61)$$

where the volume, v_0, and the isothermal compressibility, β_T, belong to the point with the coordinates $T = 0$ K, $p = 0$. It follows from Eq. (3.61) that, in order to approach the spinodal state at $T = 0$ K, the crystal volume has to be increased by approximately 30 % in comparison with the volume of the crystal at normal pressure and $T = 0$ K.

In Fig. 3.19 in the phase diagram of argon the crystal state spinodal $p_{S,sp}(T)$ is shown, calculated by using Eq. (3.60). In the temperature range from 0 K up to the melting point the data for the crystal state at atmospheric pressure were employed. Above the temperature T_0 the values $v_0 = v_S$, α_p and β_T on the melting line were taken from Ref. [37]. In the latter case the relationships for the temperature dependences of the quantities A and B have the following form

$$A = \frac{v_S^5}{2\beta_T}(1 + 3T\alpha_p) - \frac{3v_S^5}{2} p_{SL} , \qquad B = \frac{v_S^3}{2\beta_T}(1 + 5T\alpha_p) - \frac{5v_S^3}{2} p_{SL} , \quad (3.62)$$

where p_{SL} is the pressure on the melting line. The limiting pressure on the spinodal of crystalline argon at $T = 0$ K is equal to -250 MPa. This value is close to the estimate $p_{S,sp}(0) = -270$ MPa, obtained by using the Tait equation Eq. (3.50). The point of intersection of the spinodal with the zero isobar corresponds to the temperature 115 K.

In Fig. 3.19 the behavior of the internal pressure $p_{i,sp}$ of crystalline argon along the spinodal, calculated by using Eq. (3.52), is also shown. The course of the curve $p_{i,sp}(T)$ indicates a gradual increase of the internal pressure along the crystal spinodal with increasing temperature.

The limiting point on the spinodal of the crystal phase at $T = 0$ K can serve as a reference point for the investigation of the thermodynamic similarity of solids, and the crystal properties $v_{S,sp}(0)$, $p_{i,sp}(0) = -p_{S,sp}(0)$ at this point can be used as characteristic scales in building reduced thermodynamic variables. The characteristic scales of pressure and volume for the crystal state of inert gases are shown in Table 3.8.

Figure 3.19 indicates that the crystal phase on the low-temperature extension of the melting line everywhere retains stability (the spinodal is located below the melting line). But for the liquid we have another asymptotic. The spinodal, going out from the critical point C,

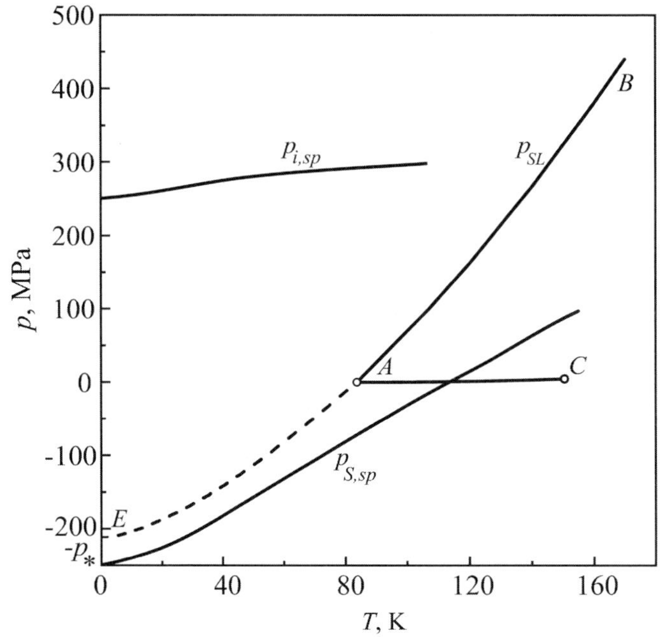

Figure 3.19: Phase diagram of argon with a region of stretched states of a crystal: (BA) is the melting line with a metastable extension (AE) into a region of negative pressures, (AC) is the liquid–vapor phase equilibrium line, $p_{S,sp}$ is the crystal spinodal, $p_{i,sp}$ is the internal pressure along the crystal spinodal.

Table 3.8: Characteristic scales of pressure and volume for the crystalline state of inert gases.

Substance	$p_{i,sp}(0)$, MPa	$v_{S,sp}(0)$, 10^{-3} m^3/kg
Neon	102	0.857
Argon	248	0.727
Krypton	315	0.417
Xenon	323	0.341

approaches the melting line in the region of negative pressures. This property is connected with a lower elasticity of the liquid as compared with the solid at a given temperature, T, and pressure, p, and confirmed by thermodynamic estimates. The constructions shown in Figs. 3.15 and 3.18 indicate the possible intersection (or touching) of the liquid spinodal and the melting line of argon. It is not correct to extend the melting line up to the limit $T = 0$ K, if the liquid phase looses stability not far from this limit. Only the crystal phase retains its stability and can exist here in a metastable state. But what then is to be done with the procedure of determination of the pressure pole p_*? This procedure ignores the limitation caused by the spinodal state of one of the phases. In the region above the triple point, where thermodynamic stability is ensured, the melting line $p_{SL} = f(T)$, corresponding to equality of the chemical potentials of the phases, is constructed without using the condition of stability. The extension of the melting line up to $T = 0$ K makes it possible to determine p_*. Such a procedure ensures agreement with experiment above the triple point, where the accuracy of the approximation can be controlled in the experiment. For temperatures lower than the triple point a limitation is introduced connected with the requirement of thermodynamic stability of the coexisting phases. A detailed specification of the behavior of the properties of the system between the point of approach of the liquid phase with the spinodal and $T = 0$ K requires a separate investigation.

In connection with these circumstances let us address once more the analogy in the determination of the ideal gas temperature scale, $pV = RT$, employing a gas thermometer, for example, a gas thermometer of constant volume. In such a case, pressure is the thermometric property at a constant mass of the gas resulting then in $p_t = p_0(1 + \alpha_V t)$. The coefficient α_V is determined by experiment taking into account the correction introduced both by the experimental device employed and the nonideality of the gas. Employing a 100 degree scale, we have $\alpha_V = 1/273.15$. The uncertainties of this determination are connected with the circumstance that the model of the ideal gas is extended to as low temperatures as is desired, while any real gas of finite density at some sufficiently low temperature will condense or reach the spinodal (supercompressed) state.

The estimate of the crystal phase spinodal, $p_{S,sp}(T)$, in the thermometric approximation and its limiting value, $p_{S,sp}(0)$, is of additional interest in connection with the problem of the ideal tensile strength of crystalline bodies [99]. For example, in Macmillan's paper [155] the maximum strain of a three-axis stretch, $\sigma_{\max} = -p_{\max} = 256$ MPa, is given as the respective value for argon at $T = 0$ K. The calculation was performed for a face-centered cubic crystal using the Lennard-Jones interatomic potential and summing over the static lattice. This value of $\sigma_{\max}$ is close to the value, $-p_{S,sp}(0) = 250$ MPa, obtained by us. In addition it should be mentioned that the pressure pole, $p_* = 211.4$ MPa, for argon is close to the discussed limit of tensile strength. The heuristic qualitative conclusion, which can be drawn from such analysis, consists in the statement that the low-temperature part of the melting line is limited with respect to pressure variations by the maximum possible tensile stretch of the condensed phase.

3.7 The Behavior of the Viscosity of a Liquid Along the Coexistence Curve with the Crystalline Phase

Considering the metastable extension of the melting curve beyond the triple point, one has to determine also the behavior of the viscosity of the liquid phase along the curve of phase equilibrium. In particular, one has to clarify whether the liquid will vitrify at $T < T_0$ or not. Vitrification would result in a significant kinetic inhibition for structural relaxation of the liquid and a delay in the establishment of liquid and crystal equilibrium at low temperatures. The analysis of experimental data shows that with the movement along the melting curve the liquid becomes more viscous with an increase of temperature and not with its decrease [139, 140, 156–158]. For some substances, the variation of the viscosity, η, along the first segment of the melting curve is shown in Fig. 3.20. The curves were plotted using experimental data for the viscosity in the immediate vicinity of the melting curves. The dependencies of viscosity on temperature along the isobars were extrapolated to the corresponding melting temperatures. In this way the values of liquid-phase viscosity on the melting curves were obtained for benzene up to 550 MPa [159], for carbon dioxide up to 300 MPa [160], for carbon tetrachloride up to 200 MPa [114] and for argon up to 780 MPa [161]. The behavior of the curves indicates that the viscosity of the liquid phase along the melting curve decreases with a decrease of temperature. If such a tendency persists, the melt may vitrify not on the low-temperature segment of the extended melting curve, but on its high-temperature part.

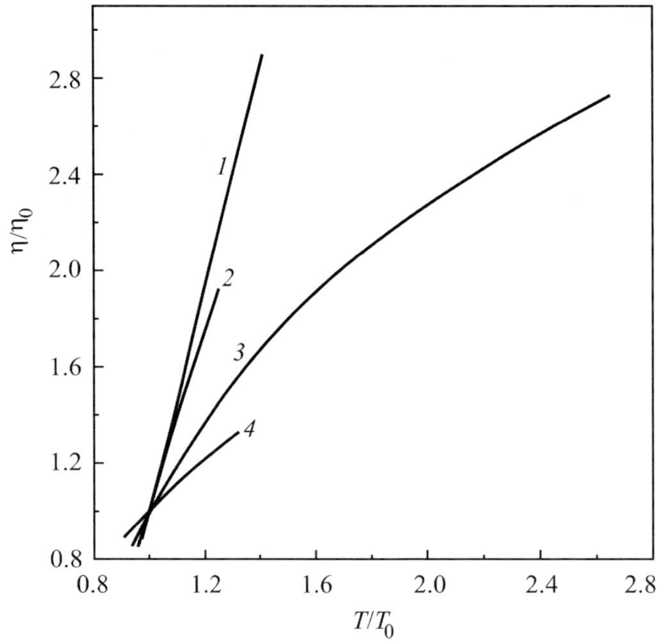

Figure 3.20: Behavior of the viscosity of several liquids along the melting curve: (1) benzene, (2) carbon dioxide, (3) argon, (4) carbon tetrachloride. Here η_0 is the viscosity of the liquid at the melting point at atmospheric pressure.

Depending on the nature of the substance under consideration, on temperature and on pressure, the viscosity of liquids can vary in a very wide range, e.g., from $\eta = 10^{-3}$ Pa·s (water at normal conditions) to 10^{12} Pa · s in the vitrification range. There is no satisfactory molecular theory of viscosity and no simple empiric correlations describing the behavior of liquid viscosity from highly fluid states to the vitrified state with a sufficiently high accuracy, although a number of such attempts have been developed in the past. According to Frenkel [17], the liquid viscosity is related to the activation energy E of the local molecular rearrangement via

$$\eta = A \exp\left(\frac{E}{k_B T}\right) , \qquad (3.63)$$

where A is a weak function of temperature and pressure in comparison with the exponential term. The activation energy increases with a decrease of temperature and an increase of pressure. With an approach to the vitrification region, the parameter E increases at atmospheric pressure by an order of magnitude in comparison to its value for low-viscosity states [6].

According to an alternative approach, the viscosity of a liquid is determined by its specific volume, v_L. This approach is related to the well-known expression proposed by Batchinsky [162]

$$\eta = \frac{C}{v_L - v_0}\,, \tag{3.64}$$

where C and v_0 are individual constants of the liquid under consideration.

The widely employed correlations

$$\eta = A \exp\left(\frac{B}{T - T_1}\right) \tag{3.65}$$

and

$$\eta = a \exp\left(\frac{b}{v_L - v_0}\right)\,, \tag{3.66}$$

known respectively as Vogel–Fulcher–Tammann [163–165] and Doolittle equations [166], should be considered as modifications of Eqs. (3.63) and (3.64). The introduction of the parameters T_1 and v_0 a priori determines the states of zero fluidity by the conditions $T = T_1$ or $v = v_0$; however, the values T_1 and v_0 depend on pressure. Moreover, in order to arrive at an agreement between experiment and theoretical predictions, the parameters A and b should be also considered as functions of pressure, while B and a can be considered as constants having specific values for the different substances under consideration.

In the present investigation the Vogel–Fulcher–Tammann correlation, Eq. (3.65), is employed. Data on the viscosity of the following liquids has been processed: argon [161], carbon tetrachloride [114], benzene [159], dodecane [167], octadecane [167], and carbon dioxide [160]. The range of variation of pressure, analyzed in these references, was 360 MPa for dodecane and 200–300 MPa for the other substances. At atmospheric pressure, the values A, B and T_1 in Eq. (3.65) were chosen taking into account that, at the temperature of vitrification, T_g, the condition $\eta = 10^{12}$ Pa · s should hold. For other isobars, the values of A and T_1 were calculated for a fixed value of B for $p = 1$ atm. The approximations

$$A = \alpha\left(\frac{p}{\beta} + 1\right)^m \tag{3.67}$$

and

$$T_1 = \gamma\left(\frac{p}{\delta} + 1\right)^n \tag{3.68}$$

were employed in order to specify the pressure dependence of the parameters $A(p)$ and $T_1(p)$, where α, β, γ, δ, m, and n are constants.

The parameters of Eqs. (3.65), (3.67) and (3.68), employed in the computations, are given in Table 3.9. For carbon dioxide and carbon tetrachloride, the dependence of the temperature parameter T_1 on pressure could not be established: for these substances the assumption

Table 3.9: The values of the parameters in Eqs. (3.65), (3.67), and (3.68) employed for the calculation of the viscosity.

Substance	B, K	α, 10^{-4} Pa·s	β, MPa	m	γ, K	δ, MPa	n
Benzene	469	0.430	182	1.068	120	403	0.677
Dodecane	394	0.860	203	1.314	150	697	0.675
Octadecane	529	0.772	243	1.828	167	1254	1.0
Argon	202	0.145	80	1.201	16	75	0.377
Carbon dioxide	724	0.045	340	2.064	36	–	–
Carbon tetrachloride	992	0.202	191	1.768	35	–	–

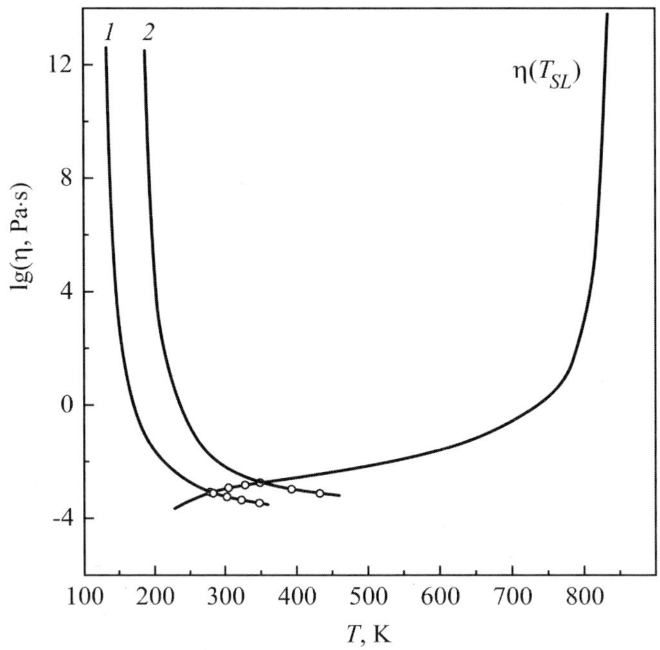

Figure 3.21: Change of the viscosity of benzene along the melting curve, $\eta\,(T_{SL})$, and along the isobars $p = 0.1$ MPa and $p = 300$ MPa (curves (1) and (2)). The circles refer to experimental data, while the full curves are computed according to Eq. (3.65).

$T_1 = \gamma$ was made. The results of the calculation of the temperature dependence of viscosity along two isobars and along the melting curve are shown for benzene in Fig. 3.21 on a semilogarithmic scale. As usual, the viscosity increases along the isobars with a decrease of temperature. The vitrification of benzene along the melting curve (corresponding to a value of the viscosity of $\eta = 10^{12}$ Pa · s) corresponds to a temperature of 833 K and a pressure of 6.5 GPa. A similar plot is shown for dodecane in Fig. 3.22. The vitrification temperature of dodecane on the melting curve equals 655 K at a pressure of 5.3 GPa.

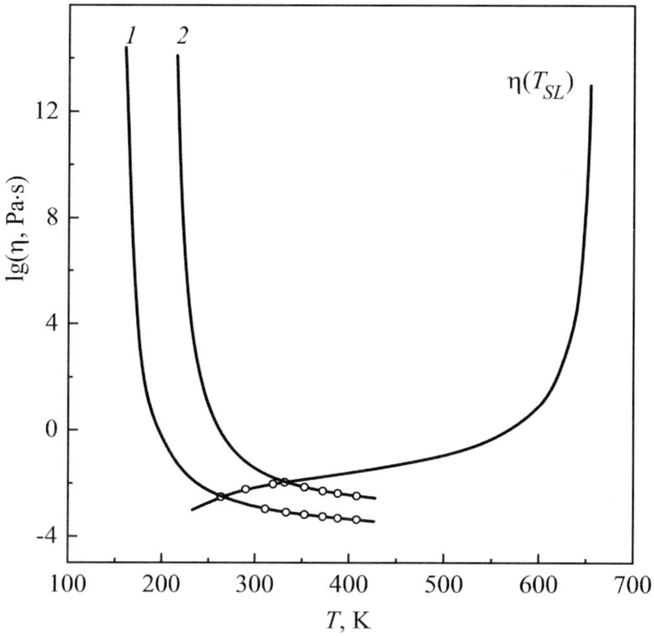

Figure 3.22: Change of the viscosity of dodecane along the melting curve, $\eta\,(T_{SL})$, and along the isobars $p = 0.1$ MPa and $p = 360$ MPa (curves (1) and (2)). The circles refer to experimental data, while the full curves are computed according to Eq. (3.65).

For liquids with a simple structure, the fluidity, $(1/\eta)$, depends significantly on the free volume, increasing with an increase of the latter. The character of the behavior of the viscosity curves of the liquid phase, as given in Fig. 3.20, allows us to suppose that the mutual influence of pressure and temperature leads to an increase of free volume of the liquid when one is moving along the melting curves towards the low-temperature range. The experiment explicitly indicates that the specific volume of the liquid increases along the melting curve with a decrease of temperature. The relative inpact of pressure and temperature with respect to the variation of the specific volume of each phase on a certain segment of the melting curve is characterized by the parameter [100, 156]

$$Ml = \frac{\beta_T}{\alpha_p}\frac{dp}{dT_{SL}}\,, \tag{3.69}$$

where $\beta_T = -(1/v)(\partial v/\partial p)_T$ is the coefficient of isothermal compressibility and $\alpha_p = (1/v)(\partial v/\partial T)_p$ is the thermal expansion coefficient. The condition $Ml > 1$ corresponds to the situation that the specific volume of the respective phase is decreased with an increase of temperature. Taking into account the known thermodynamic correlation

$$\left(\frac{\partial v}{\partial p}\right)_T \left(\frac{\partial p}{\partial T}\right)_v \left(\frac{\partial T}{\partial v}\right)_p = -1\,, \tag{3.70}$$

one obtains from Eq. (3.69) the relation

$$Ml = \frac{dp/dT_{SL}}{(\partial p/\partial T)_v} \; . \tag{3.71}$$

This result means that the inequality $Ml > 1$ holds true when the slope of the melting curve on the (p, T)-plane is larger than the slope of the isochore of the liquid (crystalline) phase at the point where the isochore intersects with the melting curve.

Table 3.10: Values of the parameter Ml of several substances for their liquid and crystal phases near the triple point.

Substance	Ml_L	Ml_S	Ref.
Hydrogen	3.0 [168]	7.0 [170]	[74]
Neon	2.4 [37]	2.2 [37]	[74]
Argon	1.9 [37]	1.6 [37]	[74]
Krypton	2.1 [37]	1.6 [37]	[74]
Xenon	2.0 [37]	1.7 [37]	[74]
Oxygen	4.3 [38]	–	[74]
Nitrogen	2.6 [39]	–	[74]
Carbon dioxide	2.9 [106]	–	[65]
Ammonia	4.0 [107]	–	[78]
Benzene	2.5 [112]	–	[113]
Lithium	15.0 [169]	12.5 [131]	[74]
Sodium	8.2 [103]	8.6 [131]	[74]
Potassium	8.8 [132]	7.2 [131]	[74]
Rubidium	6.9 [134]	7.1 [131]	[74]
Cesium	8.2 [135]	8.5 [131]	[74]
Mercury	4.1 [136]	3.9 [136]	[74]

Table 3.10 gives the values of Ml for liquid and crystalline phases calculated for several substances near their melting temperatures at atmospheric pressure using the literature data for the thermodynamic coefficients α_p, β_T, and the slope of the melting curve, dp/dT_{SL}. For most of the substances, the values of α_p and β_T were computed using (p, v, T)-data near the triple point (the references to the sources are also given in the table) or are taken from a reference book [131]. The values of β_T for liquid lithium, rubidium, and cesium were determined by extrapolating up to the corresponding temperatures the data for the compressibility, which were calculated using the results of measuring the speed of sound in the melts. A separate column contains the references to the literature sources for the derivative dp/dT_{SL}. If we suppose that the condition $Ml_L > 1$ is retained on the whole melting curve, including its low-temperature extension $T < T_0$ into the region of stretched (metastable) states, then a vitrification of the liquid on this low-temperature extension is highly improbable. The behavior of the parameter Ml_L on the melting curve of argon, shown in Fig. 3.23, may serve as a good example illustrating this conclusion. The values of Ml_L were calculated using experimental data for α_p, β_T, and dp/dT_{SL} taken from Ref. [171]. The dashed curve on the figure indicates the extension of the dependence of Ml_L on temperature into the range of negative pressures.

As follows from the figure, the parameter Ml_L is, for argon, almost constant along the melting curvc.

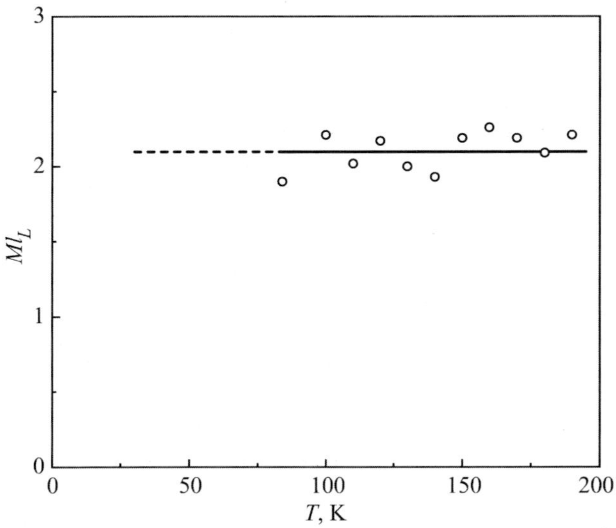

Figure 3.23: Behavior of the complex parameter Ml_L along the melting curve of argon.

One may derive also another local criterion proving that the inequality $d\eta/dT_{SL} > 0$ has to be fulfilled. Indeed, we may write [172]

$$\frac{dp/dT_{SL}}{(\partial p/\partial T)_\eta} > 1 \,. \tag{3.72}$$

Inequality Eq. (3.72) correlates, in terms of the variables T and p, the slope of the melting curve and the slope of the curve of constant viscosity. Indeed, considering the viscosity as a single-valued function of temperature and pressure, we have the following dependence connecting the partial derivatives

$$\left(\frac{\partial \eta}{\partial T}\right)_p \left(\frac{\partial T}{\partial p}\right)_\eta \left(\frac{\partial p}{\partial \eta}\right)_T = -1 \,. \tag{3.73}$$

From Eq. (3.73) and the inequalities $(\partial\eta/\partial T)_p < 0$ and $(\partial\eta/\partial p)_T > 0$ it follows that the derivative $(\partial p/\partial T)_\eta$ is positive. The total derivative, $d\eta/dT_{SL}$, of viscosity with respect to temperature along the melting curve can be expressed thus in the following form

$$\frac{d\eta}{dT_{SL}} = \left(\frac{\partial \eta}{\partial T}\right)_p + \left(\frac{\partial \eta}{\partial p}\right)_T \left(\frac{dp}{dT_{SL}}\right) \,. \tag{3.74}$$

If $(dp/dT_{SL}) > 0$ holds then both terms on the right-hand side of Eq. (3.74) have opposite signs. If we express the derivative $(\partial\eta/\partial p)_T$ via Eq. (3.74) and substitute it into Eq. (3.73), we get

$$\frac{d\eta}{dT_{SL}} = \left(\frac{\partial\eta}{\partial T}\right)_p \left[1 - \frac{dp/dT_{SL}}{(\partial p/\partial T)_\eta}\right] . \tag{3.75}$$

Taking into account the facts mentioned above one gets from latter inequality the condition Eq. (3.72).

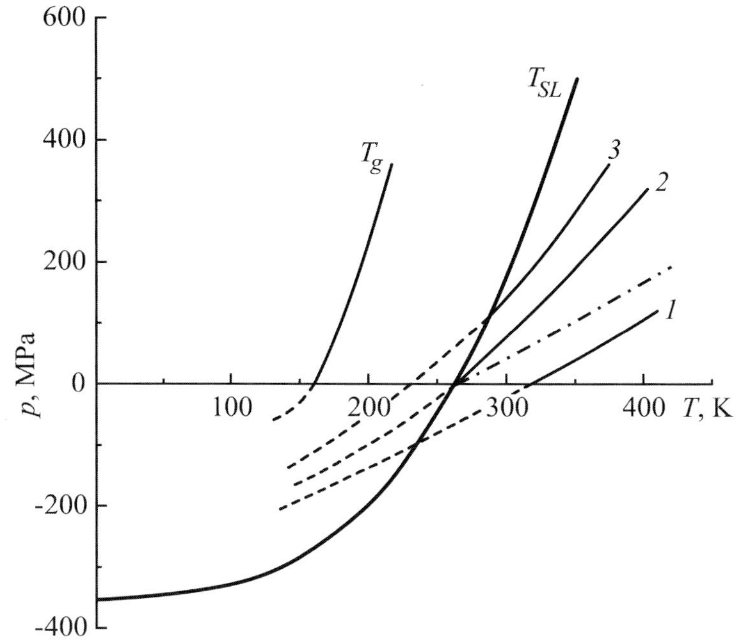

Figure 3.24: The melting curve, $T_{SL}(p)$, of dodecane, the curves of constant viscosity ((1) 1 mPa·s, (2) 3 mPa·s, (3) 5 mPa·s), and the curve corresponding to the temperature of vitrification (T_g, $\eta(T_g) = 10^{12}$ Pa·s) including its continuation into the range of negative pressures. The dashed-dotted curves refer to the isochore of the liquid with a specific volume of the liquid of $v = 1.3\,\text{cm}^3/\text{g}$.

Figure 3.24 shows, for dodecane as an example, the melting curve, $T_{SL}(p)$, the curve of vitrification of the liquid, $T_g(p)$, the isoviscosity curves 1–3 and the isochore having a common point of attachment to the melting curve with the curve $\eta\,(T,p) = 3\ \text{mPa}\cdot\text{s}$. Qualitatively, the behavior of the liquid is of interest not only by the increase of η with increase of temperature, but also by the property that low values of η are retained on the metastable extension of the melting curve into the range of low temperatures, i.e., at $T/T_0 < 1$. This result follows from the monotonic character of the iso-viscosity curves, $\eta(T,p) =$ constant, and the fact that they do not intersect with each other (cf. Fig. 3.24).

For several substances, Table 3.11 gives the melting temperature T_0, the viscosity of the respective liquid at the temperature, T_0, and atmospheric pressure and also the temperature of vitrification, T_g, at this pressure. Furthermore, the calculated values of the viscosity of the melt at $T/T_0 = 0.8$ and $T/T_0 = 1.5$ are shown as well as the pressure at melt vitrification, $(p_g)_{SL}$, on the melting curve. For one and the same substance, the data for T_g at $p = 0.1$ MPa differ significantly if the results of different authors are compared. For carbon tetrachloride, we did choose the value $T_g = 61$ K taken from Ref. [173]. In this paper, the vitrification point has been estimated using the experiments on crystallization of amorphous layers. The authors of Ref. [174] report a different value $T_g = 129$ K, which has been obtained by extrapolation of vitrification temperatures of a solution of carbon tetrachloride and ethylbenzene to the zero concentration of the second component. For $T_g = 61$ K, the value of the ratio T_g/T_0 is 0.25; this ratio is significantly lower than the corresponding value for organic substances with a more complex molecular configuration. Thus, for dichloroethane we have $T_g/T_0 = 0.40$ [6], for thiophene $T_g/T_0 = 0.39$ [6], and for benzene $T_g/T_0 = 0.47$ [174]. The values of T_g, given in the table for argon and carbon dioxide, are obtained based on the assumption that $T_g/T_0 = 0.25$ holds. For dodecane and octadecane, the value of this ratio is assumed to be equal to 0.6.

Table 3.11: Viscosity of several liquids in various points of the melting curve, vitrification temperature, T_g, at atmospheric pressure and the pressure $(p_g)_{SL}$ of vitrification of the liquid on the melting curve: (1) carbon tetrachloride, (2) benzene, (3) dodecane, (4) octadecane, (5) carbon dioxide, (6) argon.

	T_0, K	p_*, MPa	η, mPa·s $T = 0.8T_0$	η, mPa·s $\eta(T = T_0)$	η, mPa·s $\eta(T = 1.5T_0)$	$(p_g)_{SL}$, GPa	T_g, K
1	250.6	292	1.8	2.0	2.7	–	61
2	278.6	360	0.1	0.82	3.2	6.5	131
3	263.7	355	0.6	2.8	14	5.3	160
4	301.0	345	1.3	4.0	55	2.7	181
5	216.6	400	0.2	0.25	0.6	–	54
6	83.8	211	0.1	0.28	0.5	–	21

From the plot of isoviscosity curves (Fig. 3.24) it can be expected that they are nonlinear in the range of negative pressures, only in this case their intersection is impossible. At high pressures ($p \approx p_*$), the curves of constant viscosity in (T, p)-coordinates are close to straight lines. This feature is proven experimentally also for the vitrification curve [175]. When considering the question of vitrification of the liquid along the melting curve, we employ the following rule. The liquid is considered as capable of vitrification along the melting curve if the extension of the vitrification curve, $\eta\,(T,p) = 10^{12}$ Pa · s, intersects the melting curve $T_{SL}(p)$ at a pressure $p/p_* < 20$. In the alternative case, the question remains still open, since the procedure employed becomes unreliable. For the substances in the table, this is the case for argon, carbon dioxide and carbon tetrachloride.

The analysis of the behavior of the viscosity of a liquid, performed here and taking into account its continuation far beyond the triple point, gives the following picture of variation of η along the melting curve. The liquid retains a high fluidity, $1/\eta$, on the low-temperature (metastable) segment of the melting curve, which is due to the uniform stretching of the coexisting phases. The viscosity of the melt increases with the increase of temperature, and the liquid can vitrify in the neighborhood of a certain point of the melting curve, $\{T, p\}_{SL,g}$. At higher temperatures there could be kinetic obstacles in the realization of phase equilibrium.

The described behavior of the viscosity has not only a heuristic value, but also a practical one, e.g., in application to metallurgy and geophysics. In Ref. [158], it is concluded, using the extrapolation of the melting curve of iron up to $p \approx 300$ GPa, that the core of the Earth exists in the vitreous state. This conclusion, which is based on an extensive extrapolation, requires an additional analysis, but the approach itself proves the relevance of further research of liquid viscosity along the melting curve in a wide range of temperature and pressure for various substances.

3.8 The Behavior of Volume and Entropy Jumps Along the Melting Curve

The analysis of experimental data on the melting entropy of substances of different nature reveals the tendency [83] that Δs_{SL} decreases with an increase of temperature. If we take into account the metastable continuation of the melting curves beyond the triple point, then Δs_{SL} has to decrease also in the range of low temperatures. Such behavior is a consequence of the Nernst theorem which implies

$$\Delta s_{SL} \to 0 \qquad \text{at} \qquad T \to 0\,. \tag{3.76}$$

Hence, with an increase of temperature, the entropy jumps along the melting curve first increasing and then, after passing a maximum, decreasing again.

Table 3.12: Comparison of experimental [143] and calculated (by Eq. (3.83)) data for the volume jump of argon along the melting curve ($T_m = 69.8$ K).

T, K	$\Delta v_{SL(exp)}$ cm^3/mol	$\Delta v_{SL(calc)}$ cm^3/mol	$\frac{\Delta v_{SL(exp)} - \Delta v_{SL(calc)}}{\Delta v_{SL(exp)}}$
83.78	3.53	3.53	0
137.0	2.40	2.42	−0.01
182.0	1.71	1.84	−0.08
188.0	1.70	1.78	−0.05
197.0	1.62	1.69	−0.04
226.0	1.41	1.45	−0.03
245.0	1.34	1.31	0.02
262.0	1.27	1.20	0.06
278.0	1.21	1.11	0.08
284.0	1.05	1.08	−0.03
288.0	1.09	1.06	0.03
290.0	1.04	1.05	−0.01
291.0	1.12	1.04	0.07
299.0	1.04	1.01	0.03
360.0	0.69	0.76	−0.10

If we express the derivative (dp/dT_{SL}) in the Clausius–Clapeyron equation according to Eq. (3.3), then we obtain

$$\Delta s_{SL} = \Delta v_{SL} \left(\frac{p + p_*}{T} \right) c \, . \tag{3.77}$$

The condition of the maximum of the entropy jump in melting,

$$\frac{\partial(\Delta s_{SL})}{\partial T} = 0 \tag{3.78}$$

then takes the form [100]

$$-\frac{\partial \ln(\Delta v_{SL})}{\partial \ln T} = c - 1 \, . \tag{3.79}$$

This correlation connects the location of the maximum of the entropy jump with the behavior of the volume jump along the melting curve. Introducing the variable $p' = p + p_*$, we get instead of Eq. (3.79) the result

$$-\frac{\partial \ln(\Delta v_{SL})}{\partial \ln p'} = c(c - 1) \, . \tag{3.80}$$

The condition given by Eq. (3.79) has to be taken into account in the choice of the relation, which determines the dependence of the volume jump on temperature. The following expression for Δv_{SL} can then be proposed [36]

$$\Delta v_{SL} = a \exp \left[- \left(\frac{T}{T_m} \right)^{c-1} \right] , \tag{3.81}$$

where T_m is the temperature of the maximum of the entropy jump. Equation (3.81) is consistent with the Simon equation. It is convenient to express the constant a in this expression through the volume jump $(\Delta v_0)_{SL}$ at zero pressure, i.e., at the temperature T_0. We get then

$$a = (\Delta v_0)_{SL} \exp \left(\frac{T_0}{T_m} \right)^{c-1} \, . \tag{3.82}$$

It is also convenient to transform Eq. (3.81) into the form

$$\ln \left(\frac{\Delta v_{SL}}{(\Delta v_0)_{SL}} \right) = \left(\frac{T_0}{T_m} \right)^{c-1} \left[1 - \left(\frac{T}{T_0} \right)^{c-1} \right] \, . \tag{3.83}$$

Table 3.12 gives an idea about the magnitude of the deviation of experimental data for Δv_{SL} from Eq. (3.83). The value of T_m is chosen to give the best fit to experimental data in a wide range of pressures, $p > 0$.

Using Eq. (1.2), the approximations Eqs. (3.3) and (3.83), we can calculate the behavior of the entropy jump on the metastable extension of the melting curves. Figure 3.25 shows

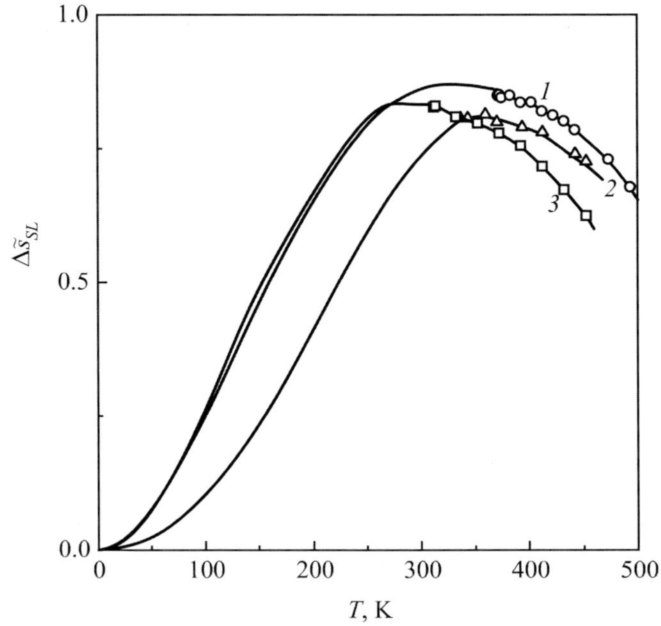

Figure 3.25: Temperature dependence of the melting entropy: (1) sodium, (2) potassium, and (3) rubidium. Symbols represent the experimental data, the full curves are the results of computations employing Eqs. (1.2), (3.3) and (3.83).

for alkali metals the dependence of the melting entropy, $\Delta\widetilde{s}_{SL}$, on temperature plotted using experimental data and their extrapolation to the segment $0 \leq T \leq T_0$ [141, 176]. The symbols show the experimental data for the melting curve and Δv_{SL} as taken from Ref. [177] for sodium and from Ref. [178] for potassium and rubidium. As follows from the figure, the melting entropy approaches its maximum with a decrease of temperature near T_0 and then rapidly decreases, again, reaching $\Delta\widetilde{s}_{SL} \to 0$ for $T \to 0$. The behavior of $\Delta\widetilde{s}_{SL}$ in the range of low temperatures qualitatively reflects the picture of the crystal–liquid phase transition and is in agreement with the requirements of the Nernst theorem for condensed phases in internal thermodynamic equilibrium.

It follows from Eq. (3.31) that the jumps of volume and entropy are closely related [100, 179]. Figure 3.26 shows the dependence between $\Delta\widetilde{v}_{SL}$ and $\Delta\widetilde{s}_{SL}$ plotted based on experimental data for sodium [177] and carbon tetrachloride [180]. The metastable segment is plotted using Eqs. (1.2), (3.3), and (3.83). The linearity is kept for the whole studied region of existence of the melting curves. If the proportionality between $\Delta\widetilde{v}_{SL}$ and $\Delta\widetilde{s}_{SL}$ is considered as an experimental fact, then the Simon equation Eq. (3.3) can be considered as a direct consequence following from it. In other words, for each substance there exists a shifted pressure scale $p' = p + p_*$ with its own pole p_*. In this representation, the exponent c in Eq. (3.3) coincides with the angular coefficient of the curve $\Delta\widetilde{s}_{SL} = c\Delta\widetilde{v}_{SL}$, which passes through the origin of the coordinates.

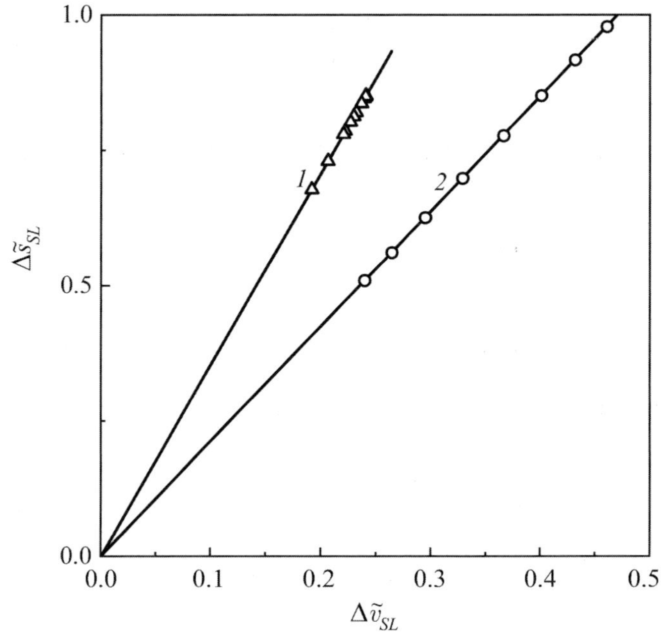

Figure 3.26: Correlation between the reduced entropy of melting and the reduced volume change in the phase transition for sodium (1) and carbon tetrachloride (2). Symbols represent the experimental data, the full curves are the results of computations employing Eqs. (1.2), (3.3) and (3.83).

3.9 The Surface Tension of Simple Liquids Along the Melting Curve

The existence of a surface density of the Gibbs free energy of the interface boundary (surface tension), σ_{SL}, is connected with the jumps of the volume properties in the interfacial layer. In contrast to the case of liquid–vapor equilibrium there does not exist any easy way to measure the surface tension of the crystal–melt interface boundary. The theoretical-computational estimates of the surface tension of the crystal–melt boundary are approximate. The computation results for σ_{SL} for metals [181] of different authors differ by an order of magnitude and do not give an impression about the dependence of σ_{SL} on temperature. As there is no critical point for crystal–liquid equilibrium, even qualitative predictions for the behavior of $\sigma_{SL}(T)$ in dependence on temperature are unreliable.

The systematic research of homogeneous nucleation in boiling [1] and crystallization [6] of substances of different nature allows us to draw the conclusion that such experiments can be employed as a source of data about the surface tension of the interfacial boundary. This task of determination of the surface tension from nucleation data has been performed systematically and comprehensively in application to nucleation of vapor bubbles in superheated liquids [1, 182]. An independent determination from nucleation data of σ_{LV} at the same temperatures using the method of capillary rise allowed us to compare the data and estimate the probable corrections to the value of the surface tension connected with the curvature of the interface boundary. Usually this correction does not exceed 2–6% in the case of radii of critical bubbles, r_*, of the order $r_* \approx 5$ nm.

Comparing the experiments on spontaneous crystallization with the theory of homogeneous nucleation one can estimate also the surface tension of the liquid–crystal interface, which is the only relevant fit parameter in Eqs. (1.16) and (1.18). For the first time this method has been implemented in Ref. [183] for organic liquids, and then in Ref. [184] for metals. The condition $G_* = 62$ for the Gibbs number of Eq. (1.17), employed in Ref. [183], corresponds to a nucleation rate $J \approx 10^6$ s^{-1}cm^{-1}.

A more detailed analysis of homogeneous nucleation of metals, water and some organic liquids has been made in a series of investigations presented in Ref. [6]. But all known experiments are performed at atmospheric and lower pressures, they give only the value of σ_{SL} at a single point of the melting curve close to $p = 0$. In contrast, the scaling value for the pressure for crystal–liquid equilibrium is p_* in Eq. (3.3). It has an order of magnitude of several GPa for metals. Using the method of thermodynamic similarity for the analysis of experimental data on the spontaneous crystallization of supercooled melts proves to be helpful for estimating the dependence $\sigma_{SL}(T)$ on the melting curve in a wide range of temperature and pressure [185, 186].

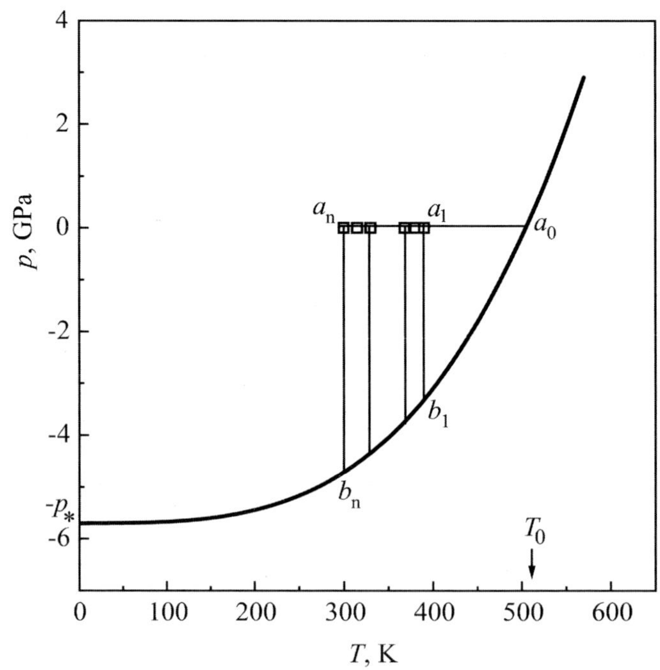

Figure 3.27: The melting curve of tin with its metastable continuation into the range of negative pressures. The data points refer to experiments on homogeneous nucleation in supercooled droplets and in island films.

Figure 3.27 shows the melting curve for tin approximated by the Simon equation Eq. (3.3) with the parameter values $T_0 = 505\,\mathrm{K}$, $p_* = 5.70\,\mathrm{GPa}$, and $c = 3.4$ [78]. It contains also the metastable extension into the region of negative pressures and low temperatures $(T/T_0) < 1$. The supercoolings $\Delta T = T_0 - T$ reached in the experiments on the kinetics of homogeneous nucleation [6] are specified by squares. The right group of points is obtained in experiments with submillimetric droplets in a neutral medium [187], the left group of points corresponds to the crystallization of tin in island films (submicrometer droplets) [188, 189]. As the result of processing the experimental data on the crystallization of supercooled melts the following two conclusions can be made [6].

1. The dependence of the nucleation rate on temperature, $J(T)$, is satisfactorily described by the theory of homogeneous nucleation with a constant value of σ_{SL} (for tin $\sigma_{SL} \approx 60\,\mathrm{mJ/m^2}$).

2. The dome-like dependence $J(T)$ is transformed into a straight line in coordinates $\lg J$ versus $[T(\Delta T)^2]^{-1}$ at $T > T_{max}$, where T_{max} is the temperature of the maximum of the function $J(T)$ (see Fig. 3.28).

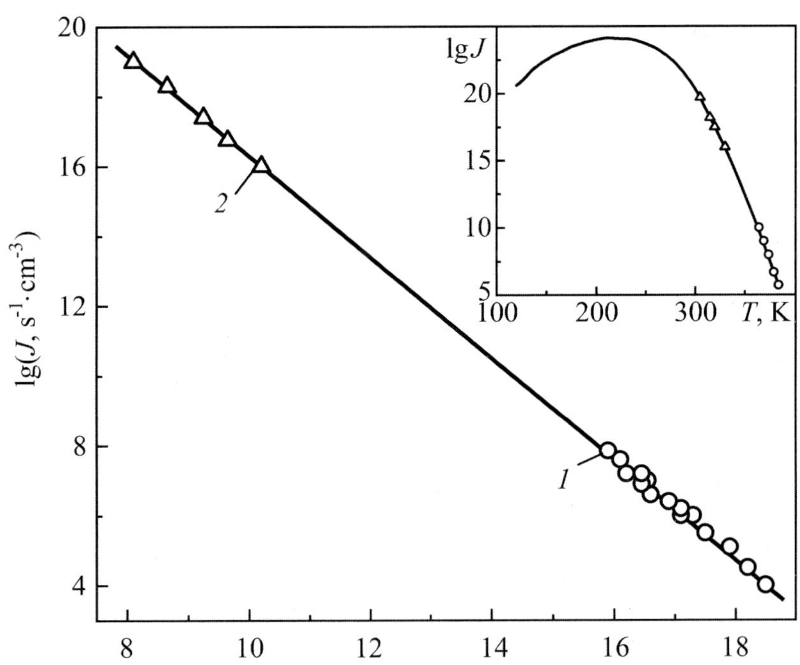

Figure 3.28: Temperature dependence of the steady-state nucleation rate, J, in supercooled liquid tin performed in different ways: (1) experiments with droplets of diameters in the range $d = 20$–$100\ \mu\mathrm{m}$, (2) experiments on droplets of island films ($d = 4$–$40\,\mathrm{nm}$). The full curve represents the results of computations by homogeneous nucleation theory.

According to Eqs. (1.16)–(1.18), this fact implies the constancy of the individual dimensional complex

$$Z = \frac{\sigma_{SL}^3 v_S^2}{k_B (\Delta s_{SL})^2}\,, \tag{3.84}$$

which is the angular coefficient for the dependence of $\lg J$ on $[T(\Delta T)^2]^{-1}$, if we assume $\ln J$ = constant $- G_*$, i.e., if we neglect the variation of $\ln(N_1 B_1)$ in comparison with the change of G_*.

Table 3.13: Values of the parameters characterizing crystal–liquid equilibrium at atmospheric pressure for a number of substances; Nc is the dimensionless complex as defined by Eq. (3.85).

Substance	T_0, K	v_S, 10^{-3} m^3/kg	Δh_{SL}, kJ/kg	$\sigma_{SL}(T_0)$, mJ/m^2	$(Nc)^{1/3}$
Lithium	453.8	1.902	428.6	30	0.44
Sodium	370.8	1.019	115.2	20	0.50
Copper	1356	0.119	203.1	200	0.54
Silver	1235	0.102	104.7	143	0.55
Indium	429.8	0.139	28.4	31	0.49
Tin	505.0	0.139	60.7	60	0.54
Mercury	234.3	0.070	11.5	23	0.51
Lead	600.0	0.091	22.6	40	0.50
Argon	83.8	0.608	29.8	7	0.5
Carbon tetrachloride	250.6	0.570	16.4	6.7	0.47
Benzene	278.6	0.990	128.2	21.7	0.54
Gallium	302.9	0.175	80.3	40	0.41
Antimony	903.7	0.152	163.9	101	0.41
Bismuth	544.5	0.103	52.6	69	0.55
Water	273.2	1.090	334.4	28.7	0.41

For p = constant, the set of experimental data $(a_1, a_2, \ldots, a_n)$ refers to a single point (a_0) of the melting curve (see Fig. 3.27). The specific volume, v_S, and the melting entropy, Δs_{SL}, are taken for this point. Then, if the condition Z = constant holds at increasing supercooling and fixed values of v_S and Δs_{SL} constancy of σ_{SL} follows as a consequence. A wider interpretation of the condition Z = constant can be given, if the points a_i are referred to the different points b_i on the melting curve as shown in Fig. 3.27, where pairs a_i, b_i correspond to the same temperature. Each point b_i is characterized by its own values of Δs_{SL} and v_S. Then, in general from the condition Z = constant we get different values of σ_{SL} for the points $b_1, b_2, \ldots$. The fact mentioned above of the transformation of the dependence $J = J(T)$ into a straight line in appropriate coordinates (cf. Fig. 3.28) gives a strong indication of the existence of a correlation between σ_{SL}, Δs_{SL} and v_S in the form of the condition Z = constant, where Z is given by Eq. (3.84). By dividing both sides of Eq. (3.84) by T_0^3 the complex Z can be brought into a dimensionless form $Nc = Z/T_0^3$,

$$Nc = \frac{\sigma_{SL}^3 v_S^2}{k T_0^3 (\Delta s_{SL})^2} = \frac{\sigma_{SL}^3 v_S^2}{k T_0 (\Delta h_{SL})^2}\,, \tag{3.85}$$

where T_0 is the temperature on the melting lines of different substances in reference points defined via the condition $p = 0$.

The analysis of experimental data leads to the conclusion that the values of $(Nc)^{1/3}$ are close for different substances. Table 3.13 contains the values defining the parameter Nc for a number of substances in reference points of the melting curves corresponding to zero (atmospheric) pressure. The values of T_0, v_S, and Δh_{SL} for lithium and sodium were taken from Ref. [77], for the others from Ref. [190]. For $\sigma_{SL}(T_0)$, data were taken from Ref. [6] except for sodium, lithium [181] and argon. For argon the value of $\sigma_{SL}(T_0)$ was obtained from the condition $(Nc)^{1/3} = 0.5$, which corresponds to the mean value of the dimensionless complex for normally melting substances.

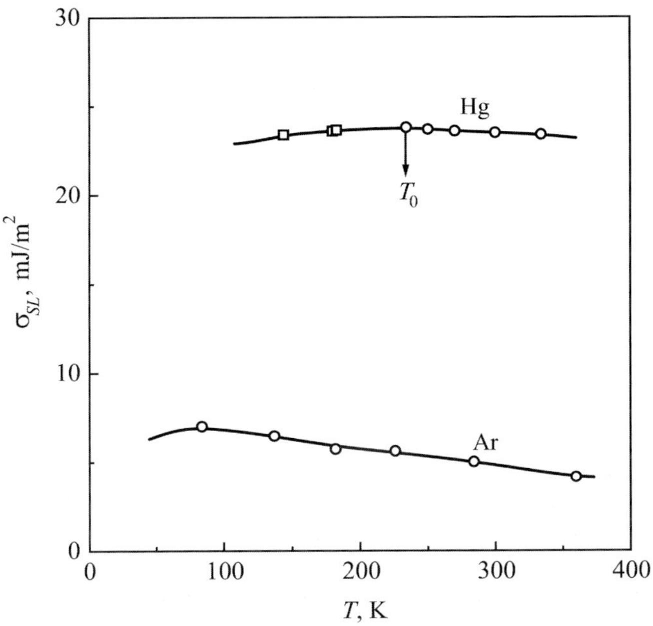

Figure 3.29: Temperature dependence of the surface tension of argon and mercury along the melting curve.

The use of the complex parameter as given by Eqs. (3.84) and (3.85) for the determination of σ_{SL} of the crystal–liquid phase boundary implies the utilization of the phenomenology of thermodynamic similarity. The surface tension is considered here as an explicit function only of temperature. The pressure corresponds to this temperature on the melting curve or on its metastable extension into the region of negative pressures, $p < 0$. Let us assume that the condition $Nc =$ constant is retained not only near $p = 0$, but also on a wide range along the melting line of each substance. Then for the calculation of the dependence $\sigma_{SL}(T)$ one has to know the entropy, Δs_{SL}, or enthalpy, $\Delta h_{SL} = T\Delta s_{SL}$, jumps and the specific volume of the crystal phase in various points of the melting line, $p_{SL} = f(T)$. The results of the computations of $\sigma_{SL}(T)$ are shown in Fig. 3.29 for mercury and argon, and in Fig. 3.30 for tin, lead and sodium. The data for Δs_{SL} and v_S were taken from the literature for values in the range

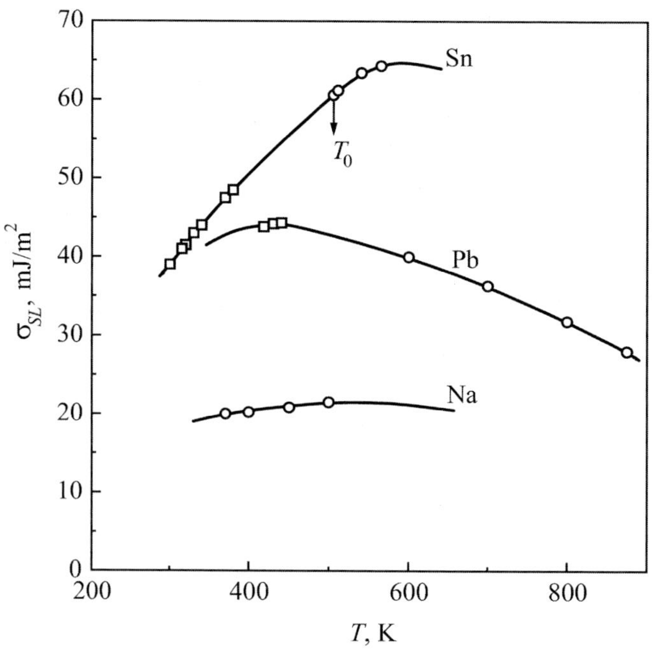

Figure 3.30: Temperature dependence of the surface tension of tin, lead and sodium along the melting curve.

$p_{SL} > 0$ and were determined for the range of pressures $p_{SL} < 0$ on the basis of approximations consistent with the Simon equation [100]. The squares specify the values of σ_{SL} in the temperature region where the nucleation kinetics was studied experimentally [6]. The circles denote the points calculated by employing the condition $Nc =$ constant and experimental data for Δh_{SL} and v_S from the literature. The smooth curves were plotted by extrapolating the values of Δh_{SL} and v_S.

For mercury and tin the temperature T_0 is shown by arrows. For the other substances, presented in Figs. 3.29 and 3.30, the temperature T_0 corresponds to the leftmost circle. The values of σ_{SL}, specified by squares, are related to negative pressures since spontaneous crystallization was observed [6] at a finite supercooling of the liquid, $T < T_0$, and the projection of the image point for the given temperature T is located on the segment of the melting line where $p < 0$ holds. These circumstances are not in contradiction with the fact that the values of $\sigma_{SL}(T_0)$ for mercury and tin were determined from the experiments on the nucleation kinetics at fixed values of Δh_{SL} and v_S, taken at T_0 (see the comments to Fig. 3.27).

The width of the temperature region at which the variation of σ_{SL} could be studied (cf. Figs. 3.29 and 3.30) depends on the availability of data for Δh_{SL} and v_S at high pressures. For argon the rightmost point corresponds to a pressure of 1.77 GPa [143], for mercury to 2.00 GPa [136], for tin to 2.60 GPa [191], for lead to 5.00 GPa [72, 192] and for sodium to 2.20 GPa [177]. The obtained variation of $\sigma_{SL}(T)$ is characterized by the existence of a maximum along the melting line. This result is connected with the properties of the ratio

$(\Delta s_{SL}/v_S)^{2/3}$ in Eq. (3.85). It is worth noting that the decrease of the entropy jump, Δs_{SL}, at $T/T_0 \ll 1$ is a consequence of the Nernst theorem. Equation (3.3) for $c > 1$ satisfies the condition $dp/dT_{SL} \rightarrow 0$ at $T \rightarrow 0$, which implies according to Eq. (1.2) that the entropy jump, Δs_{SL}, tends to zero at $T \rightarrow 0$ and finite values of the volume difference, Δv_{SL}. The decrease of σ_{SL} with the decrease of temperature corresponds to the approach to the spinodal state of crystal and liquid on the melting line with their stretching. The existence of coexisting phases at $T \rightarrow 0$ K presumes the preservation of a finite value of σ_{SL} at $T = 0$. At high temperatures, $(T > T_0)$, the derivative $(d\sigma_{SL}/dT)$ is negative due to the increase of the amplitude of atomic oscillations, which leads to the swelling of the phase boundary and to a decrease of σ_{SL}.

3.10 Correlations Between Thermodynamic Properties Characterizing Melting

The application of the concepts of thermodynamic similarity of the melting curves leads to the discovery of a number of universal correlations which connect the generalized thermodynamic variables characterizing the liquid–crystal phase coexistence [100, 141, 193]. These correlations reflect the thermodynamic similarity of substances with respect to the melting process and turn out to be useful for the calculation of the unknown melting curves of a substance employing a minimal set of data concerning melting at atmospheric pressure. Let us consider the "ideal" case of a phase transition defined by the conditions $\Delta u_{SL} = 0$ and $\Delta v_{SL} \neq 0$. As an example, one can take a phase transition in a system of solid spheres [194]. In this case, according to Eqs. (3.14), (3.18), and (3.24), we have $p_* = 0$, $p' = p$, $\Delta s_{SL}/\Delta v_{SL} = p/T$, $c = 1$, and $(\Delta\widetilde{s}_0)_{SL} = \Delta\widetilde{v}_{SL}$. Let us check, now, whether it is useful to express in the general case the term $\Delta\widetilde{v}_{SL}$ via Eq. (3.31) as

$$\Delta\widetilde{s}_{SL} = c\Delta\widetilde{v}_{SL} = \Delta\widetilde{v}_{SL} + (\Delta\widetilde{s}_1)_{SL} \; . \tag{3.86}$$

Here we obtain the following expression for the second component of the entropy jump

$$(\Delta\widetilde{s}_1)_{SL} = (c-1)\Delta\widetilde{v}_{SL} \; . \tag{3.87}$$

When c is kept constant, the values of $(\Delta\widetilde{s}_1)_{SL}$ are shifted with respect to $\Delta\widetilde{s}_{SL}$ by the factor $(c-1)/c$, the positions of their extrema coincide. Figure 3.31 shows for sodium and carbon tetrachloride the calculated dependencies of $\Delta\widetilde{s}_{SL}$ and $(\Delta\widetilde{s}_1)_{SL}$ on temperature. The points specify the experimental data for the melting line, the values of Δv_{SL} for sodium are taken from Refs. [74, 177] and from Ref. [180] for carbon tetrachloride.

The calculations performed for a large number of simple substances [100,141,193], where the melting is reduced to a positional disordering of the crystal structure, have shown that at atmospheric pressure the value $(\Delta\widetilde{s}_1)_{SL}$ is almost constant for all these substances. For the first time, this fact was noticed by Skripov in 1984 [195]. The average value of $(\Delta\widetilde{s}_1)_{SL}$ is close to 0.61. The correlation of the values $\Delta\widetilde{v}_{SL}$ and $c-1$ can be observed from Fig. 3.32 where the plot is given in coordinates $(\Delta\widetilde{v}_{SL})^{-1}$ and $c-1$. The slope of the straight line is defined by $(\Delta\widetilde{s}_1)_{SL} = 0.61$. The mean square deviation of the array of points from the linear dependence Eq. (3.87) is 0.15. The values of $(\Delta\widetilde{s}_1)_{SL}$ for some simple substances are given

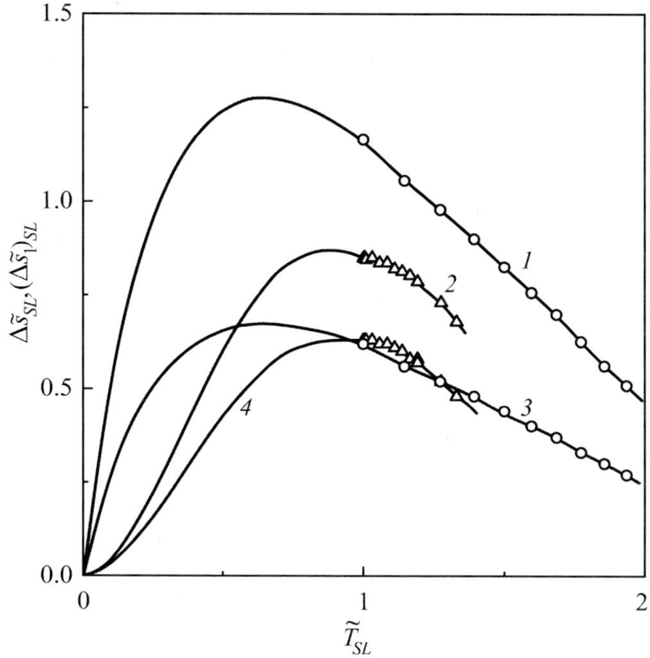

Figure 3.31: Dependence of the quantities $\Delta\widetilde{s}_{SL}$ (curves (1) and (2)) and $(\Delta\widetilde{s}_1)_{SL}$ (curves (3) and (4)) on reduced temperature for sodium (2, 4) and carbon tetrachloride (1, 3). Symbols represent the experimental data, the full curves are the results of computations employing Eqs. (1.2), (3.3) and (3.83).

in Table 3.14. This table also contains the values of Δv_{SL} which were used for the calculation with the parameters c and p_* given in Table 3.6. With an acceptable precision, the obtained values of $(\Delta\widetilde{s}_1)_{SL}$ may be referred to zero pressure (as compared with the scaling pressure $p_* \approx 0.1$–1 GPa), i.e., as to the reference points of different substances.

The approximate constancy of the value $(\Delta\widetilde{s}_1)_{SL}$ in the reference points of the melting curve of similarly melting substances reflects the universal character of the phase transition. It is revealed when the "ideal" part of the entropy jump $(\Delta\widetilde{s}_0)_{SL} = \Delta\widetilde{v}_{SL}$ is subtracted from the whole value. This result is obtained in the approximation of the Simon equation, since the explicit expression for $(\Delta\widetilde{s}_1)_{SL}$ contains the parameters c and p_*. We have

$$(\Delta\widetilde{s}_1)_{SL} = (c-1)\frac{\Delta v_{SL}(p+p_*)}{RT} \,. \tag{3.88}$$

Note, however, that the splitting of $\Delta\widetilde{s}_{SL}$ into two components

$$\Delta\widetilde{s}_{SL} = \Delta\widetilde{v}_{SL} + (\Delta\widetilde{s}_1)_{SL} \tag{3.89}$$

may also be obtained without using Eq. (3.3). It is sufficient to postulate the existence of a characteristic individual internal pressure, p_*, which is included by the definition Eq. (3.30)

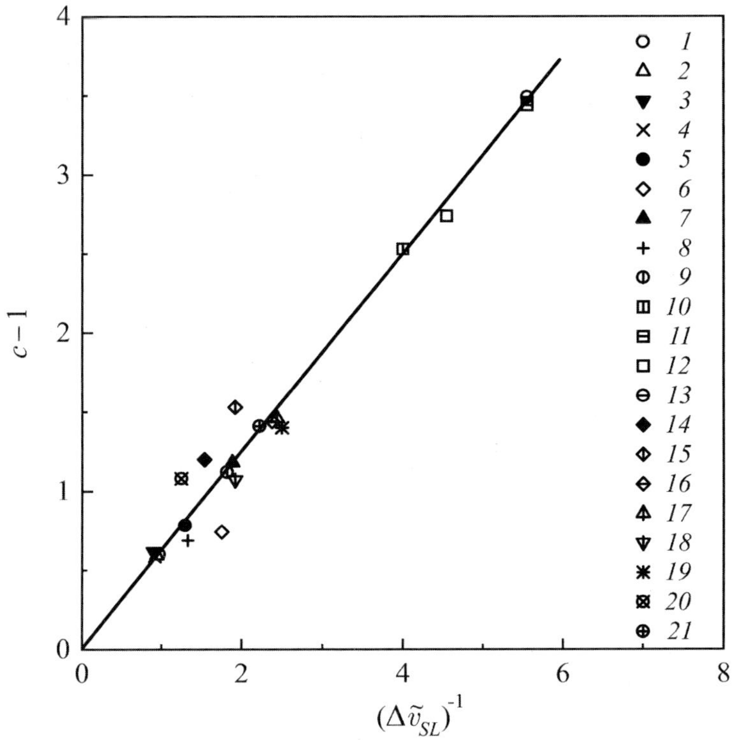

Figure 3.32: Correlation between the parameters $(c-1)$ and $(\Delta\widetilde{v}_{SL}^{-1})$ for a group of simple substances at atmospheric pressure: (1) neon, (2) argon, (3) krypton, (4) xenon, (5) nitrogen, (6) oxygen, (7) fluorine, (8) methane, (9) carbon tetrachloride, (10) sodium, (11) potassium, (12) rubidium, (13) cesium, (14) aluminum, (15) nickel, (16) copper, (17) zinc, (18) silver, (19) cadmium, (20) gold, (21) lead.

in $\Delta\widetilde{v}_{SL}$ as

$$\Delta\widetilde{v}_{SL} = \Delta v_{SL}\frac{(p+p_*)}{RT} \; . \tag{3.90}$$

It would also be interesting to compare the values of $(\Delta\widetilde{s}_1)_{SL}$ for different substances at other reference points, not just at $\widetilde{p}_{SL} = p'/p_* = 1$, as is done in Table 3.14 and Fig. 3.32, but such a procedure would require experimental data for the volume jump, Δv_{SL}. Unfortunately, for most of the substances such data are available only for a pressure close to atmospheric.

Useful information may be obtained by considering the melting line and the system of isochores for liquid and crystalline states on a (T,p)-plane. Through each of the points of the melting line, one liquid and one "crystal" isochore pass. Let us write down in a differential form the equation for the liquid along the coexistence curve with the crystal. We get

$$\frac{dp}{dT_{SL}} = \left(\frac{\partial p}{\partial T}\right)_v + \left(\frac{\partial p}{\partial v}\right)_T \frac{dv_L}{dT_{SL}} . \tag{3.91}$$

Table 3.14: Volume and entropy jumps at melting of simple substances at atmospheric pressure. Here $(\Delta\widetilde{s}_1)_{SL} = \Delta\widetilde{s}_{SL} - \Delta\widetilde{v}_{SL}$ holds and $\Delta\widetilde{s}_h$ is the variation of the entropy at a homophase expansion of the liquid by Δv_{SL}. Finally, $\Delta\widetilde{s}_{SL} - \Delta\widetilde{s}_h$ is the variation of entropy due to the disordering of the crystal structure.

Substance	Δv_{SL} $\mathrm{cm^3/mol}$	$\frac{\Delta v_{SL}}{v_S}$	$\Delta\widetilde{v}_{SL}$	$\Delta\widetilde{s}_{SL}$	$(\Delta\widetilde{s}_1)_{SL}$	$\Delta\widetilde{s}_h$	$\Delta\widetilde{s}_{SL} - \Delta\widetilde{s}_h$	Ref.
Neon	2.03 [74]	0.156	1.03	1.65	0.62	0.80	0.85	[37]
Argon	3.50 [143]	0.142	1.06	1.69	0.63	0.97	0.72	[171]
Krypton	4.48 [143]	0.151	1.11	1.80	0.69	1.01	0.79	[37]
Xenon	5.59 [143]	0.151	1.09	1.73	0.64	0.94	0.79	[37]
Nitrogen	2.50 [105]	0.084	0.77	1.37	0.60	0.71	0.66	[39]
Oxygen	0.94 [74]	0.075	0.57	0.99	0.42	0.46	0.53	[38]
Fluorine	0.94 [74]	0.044	0.53	1.16	0.63	–	–	–
Methane	2.66 [105]	0.081	0.75	1.26	0.51	0.66	0.60	[42]
Carbon tetrachloride	3.92 [180]	0.053	0.55	1.17	0.62	–	–	–
Sodium	0.640 [196]	0.0264	0.25	0.88	0.63	0.11	0.77	[103]
Potassium	1.173 [178]	0.0254	0.18	0.80	0.62	0.09	0.71	[132]
Rubidium	1.450 [178]	0.0274	0.22	0.83	0.61	0.12	0.71	[132]
Cesium	1.691 [197]	0.0243	0.18	0.81	0.63	0.08	0.73	[132]
Aluminum	0.724 [74]	0.064	0.65	1.43	0.78	0.52	0.91	–
Nickel	0.45 [74]	0.063	0.52	1.32	0.80	0.59	0.73	–
Copper	0.350 [74]	0.046	0.42	1.02	0.60	0.31	0.71	–
Zinc	0.39 [74]	0.041	0.41	0.98	0.57	0.26	0.72	–
Silver	0.574 [74]	0.052	0.52	1.07	0.55	0.45	0.62	–
Cadmium	0.44 [74]	0.033	0.40	0.96	0.56	0.27	0.69	–
Gold	0.55 [74]	0.055	0.80	1.17	0.37	0.49	0.68	–
Lead	0.70 [74]	0.037	0.45	1.08	0.63	0.32	0.76	–

This dependence correlates the angular coefficients of the melting line, dp/dT_{SL}, and the corresponding liquid isochore, $(\partial p/\partial T)_v$, in their common point. For normally melting substances, we have $(dp/dT_{SL}) > (\partial p/\partial T)_v > 0$ since both factors in the last term in Eq. (3.91) are negative. Substituting in the last inequalities the derivative (dp/dT_{SL}) by the ratio $(\Delta s_{SL}/\Delta v_{SL})$, and the derivative $(\partial p/\partial T)_v$ by $(\partial s/\partial v)_T$, we obtain

$$\frac{\Delta s_{SL}}{\Delta v_{SL}} > \left(\frac{\partial s}{\partial v}\right)_T \qquad \text{or} \qquad \Delta s_{SL} > \left(\frac{\partial s}{\partial v}\right)_T \Delta v_{SL} \,. \tag{3.92}$$

These inequalities reflect the fact that the jump of entropy at the phase transition is larger than at a homophase change of the volume by the same value, Δv_{SL}. We denote the value on the right-hand side of the last inequality by Δs_h if it is defined for the homophase expansion of the liquid. Employing the inequality

$$\left(\frac{\partial p}{\partial T}\right)_v = \left(\frac{\partial s}{\partial v}\right)_T , \tag{3.93}$$

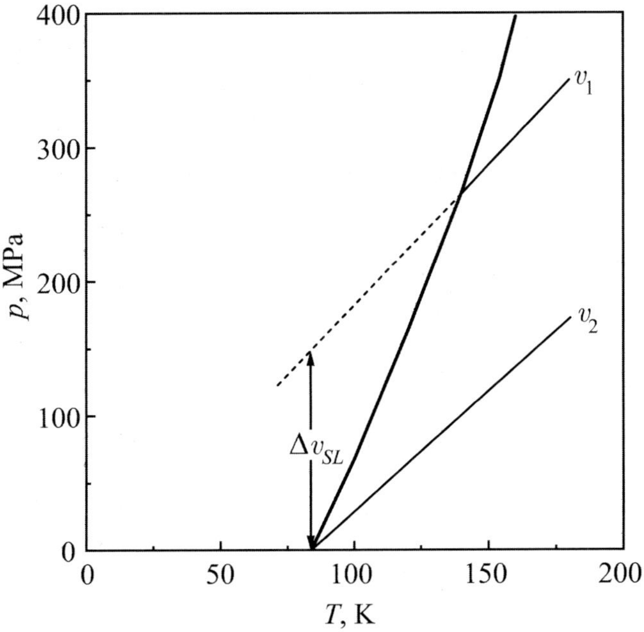

Figure 3.33: Melting curve of argon. v_1 and v_2 are two isochores of the liquid corresponding to $v_1 = 24.71$ cm^3/mol and $v_2 = 28.27$ cm^3/mol. The value v_1 corresponds to the specific volume of the crystal phase at $T_0 = 83.8$ K.

we may write further

$$\Delta s_h = \int_{v_1}^{v_2} \left(\frac{\partial p}{\partial T} \right)_v dv \;, \tag{3.94}$$

where the notation $v_2 - v_1 = v_L - v_S = \Delta v_{SL}$ is used and Δv_{SL} is the volume jump at the given temperature of crystal–liquid equilibrium, e.g., at T_0 (cf. Fig. 3.33).

Having at one's disposal the (p, T, v)-data, allowing an extrapolation into the region of metastable (supercooled) states, one can calculate the value of Δs_h at a given temperature. Using this approach for argon at $T_0 = 83.8$ K, the results $\Delta s_h = 8.06$ J/(K·mol) and $\Delta \widetilde{s}_h = 0.97$ [198] have been obtained. The difference between the reduced entropy jump at melting of argon, $\Delta \widetilde{s}_{SL} = 1.69$, and $\Delta \widetilde{s}_h$ is 0.72. This excess characterizes the melting process connected with the change of the molecular structure of the system. For argon, the parameter $\Delta \widetilde{s}_{SL} - \Delta \widetilde{s}_h$ turns out to be larger than $(\Delta \widetilde{s}_1)_{SL} = 0.63$, but is close to it. For $T \cong T_0$, the estimate $\Delta \widetilde{s}_V \approx 0.63$ of the part of the melting entropy which does not depend on volume, has been derived in Ref. [199]. In that paper, this parameter is connected with the disappearance of the transversal phonon states at melting of the crystal. An estimate of the homophase increase of the entropy of the liquid, Δs_h, is made also for other substances [198] with different degrees of accuracy at reference points on the melting line defined by $\widetilde{T}_{SL} = 1$.

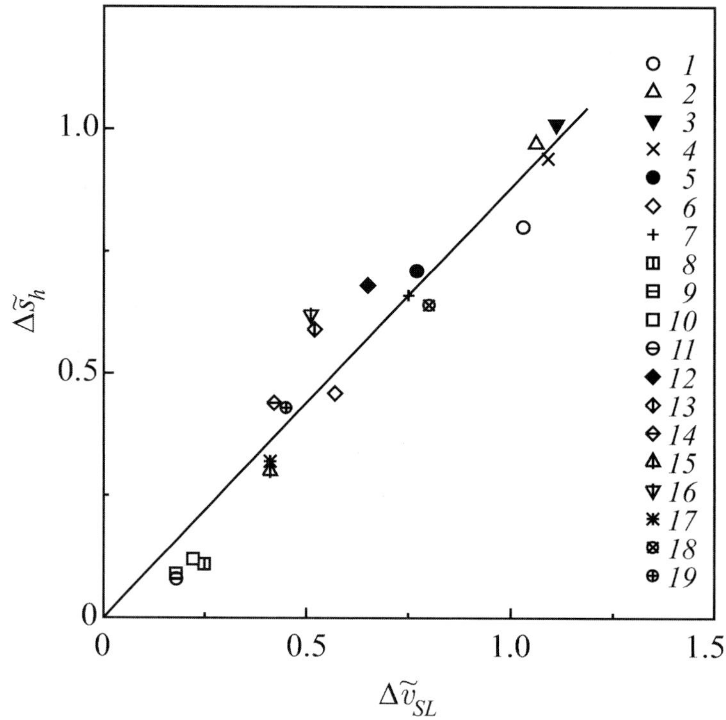

Figure 3.34: Correlation between reduced volume jump at melting and reduced homophase change of entropy for a group of simple substances: (1) neon, (2) argon, (3) krypton, (4) xenon, (5) nitrogen, (6) oxygen, (7) methane, (8) sodium, (9) potassium, (10) rubidium, (11) cesium, (12) aluminum, (13) nickel, (14) copper, (15) zinc, (16) silver, (17) cadmium, (18) gold, (19) lead. The slope of the full line corresponds to $\alpha = 0.9$.

The calculated values of $\Delta\widetilde{s}_h$ and $\Delta\widetilde{s}_{SL} - \Delta\widetilde{s}_h$ are given in Table 3.14. A separate column in this table contains the references to the literature from where the (p, v, T)-properties for the calculation of these parameters were taken. For most of the substances, where detailed (p, v, T)-data are available in a wide range of pressure variations, Eq. (3.94) has been used in the computations. For alkali metals, the value Δs_h has been determined employing the following approximation

$$\Delta s_h = \left(\frac{\partial p}{\partial T}\right)_{v=v_2} \Delta v \; . \tag{3.95}$$

In this case it was assumed that the change of the slope of the isochores with a variation of the density of the liquid is insignificant. For the other metals in Table 3.14, starting with aluminum, the quantity $\Delta\widetilde{s}_h$ refers to a homophase expansion of the crystal phase at the melting point by a value Δv_{SL}. The derivative $(\partial p/\partial T)_v$ in Eq. (3.95) is determined based on data for the coefficient α_p taken from Ref. [131] and the values of β_T are computed based on data about the elastic properties of crystals given in Ref. [133]. It is supposed in the computations

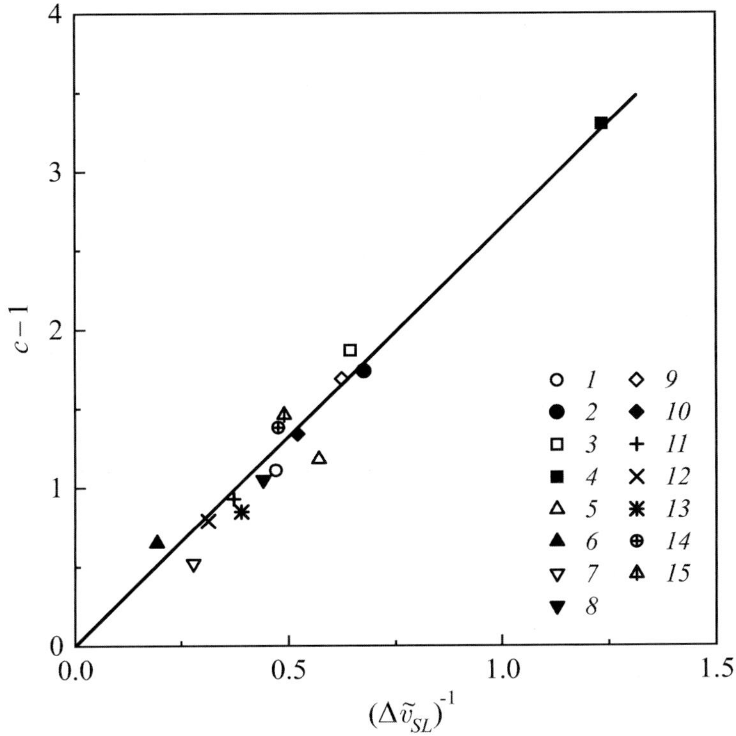

Figure 3.35: Correlation between the parameters $(c - 1)$ and $(\Delta\widetilde{v}_{SL})^{-1}$ for several molecular substances: (1) chlorine, (2) iodine, (3) carbon dioxide, (4) ammonia, (5) ethane, (6) pentane, (7) chloroform, (8) bromoform, (9) benzene, (10) aniline, (11) nitrobenzene, (12) monofluorobenzene, (13) hexafluorobenzene, (14) monochlorobenzene, (15) monobromobenzene.

that the change of the entropy at a homophase isothermal expansion of the crystal at the melting point and the homophase compression of the liquid by a value Δv_{SL} are the same. In all cases shown in Table 3.14, the difference $\Delta\widetilde{s}_{SL} - \Delta\widetilde{s}_h$ and the parameter $(\Delta\widetilde{s}_1)_{SL}$ are of comparable magnitude. The excess of $\Delta\widetilde{s}_{SL}$ as compared with $\Delta\widetilde{s}_h$ results from the additional contribution to the entropy connected with the disordering of the systems in the melting process. This contribution is absent in the isothermal homophase change of the volume by the value $\Delta v = v_L - v_S$.

Equations (3.34) and (3.95) allow us to conclude that at a homophase expansion of the crystal by a value Δv at the melting point at $p = 0$, the pressure

$$p_h = \frac{T_0 \Delta s_h}{\Delta v} \tag{3.96}$$

has values near to the internal pressure in the crystal, $p_{i,S}(T_0)$. For the case of a homophase expansion of the crystal at the melting point by a value equal to the volume jump at melting,

Δv_{SL}, one may write the following relation

$$p_{i,S}(T_0) \cong \frac{T_0 \Delta s_h}{\Delta v_{SL}} . \tag{3.97}$$

This equation connects the internal pressure with the entropy of the homophase expansion and the thermal energy $T_0 \Delta s_h$, required for an increase of the volume of the crystal by the value Δv_{SL}.

Going over to dimensionless variables according to Eq. (3.30), taking into account the correlation Eq. (3.44) between the pressure pole, p_*, and the internal pressure, $p_i(T_0)$, of the condensed phase, we arrive at the following relation between the reduced volume jump at melting and the reduced entropy change of the condensed phase in the melting point

$$\Delta \widetilde{s}_h = \alpha \Delta \widetilde{v}_{SL} . \tag{3.98}$$

The correlation Eq. (3.98), shown for a group of simple substances in Fig. 3.34, verifies that the quantities $\Delta \widetilde{s}_h$ and $\Delta \widetilde{v}_{SL}$ have similar values.

Table 3.15: Volume and entropy jumps at atmospheric pressure for substances undergoing at melting an orientational disordering of the molecules.

Substance	Δv_{SL}, cm^3/mol	$\Delta \widetilde{v}_{SL}$	$\Delta \widetilde{s}_{SL}$	$(\Delta \widetilde{s}_1)_{SL}$	$\Delta \widetilde{s}_h$	$\Delta \widetilde{s}_{SL} - \Delta \widetilde{s}_h$	Ref.
Chlorine	6.14 [74]	2.12	4.47	2.35	–	–	–
Iodine	10.20 [74]	1.48	4.06	2.58	–	–	–
Carbon dioxide	8.28 [106]	1.55	4.65	3.10	1.90	2.75	[106]
Ammonia	2.5 [59]	0.81	3.48	2.67	0.75	2.73	[107]
Ethane	5.16 [59]	1.75	3.81	2.06	–	–	–
Pentane	9.14 [111]	5.2	8.57	3.37	–	–	–
Chloroform	7.51 [204]	3.58	5.44	1.86	–	–	–
Bromoform	9.88 [204]	2.26	4.63	2.37	–	–	–
Benzene	10.3 [113]	1.60	4.30	2.70	1.56	2.74	[112]
Aniline	7.95 [205]	1.92	4.49	2.57	–	–	–
Nitrobenzene	10.0 [113]	2.68	5.17	2.49	–	–	–
Monofluorobenzene	10.2 [193]	3.19	5.71	2.52	–	–	–
Hexaflurorobenzene	13.0 [193]	2.56	4.74	2.18	–	–	–
Monochlorobenzene	8.0 [204]	2.10	5.08	2.98	–	–	–
Monobromobenzene	8.86 [204]	2.11	5.19	3.08	–	–	–

The splitting of the reduced entropy into two parts (according to Eq. (3.86)) takes into account two contributions to the entropy jump. The first one is connected with the reduced volume jump, $\Delta \widetilde{v}_{SL}$. At $\widetilde{T}_{SL} = 1$, one has $\Delta \widetilde{v}_{SL} = \Delta v_{SL} p_* / (RT_0)$; the reduction is made using the shifted pressure scale Eq. (3.19). The second contribution, $(\Delta \widetilde{s}_1)_{SL} \cong 0.61$, reflects universal features of the melting mechanism in the presence of interacting particles (it is absent in a system of solid spheres).

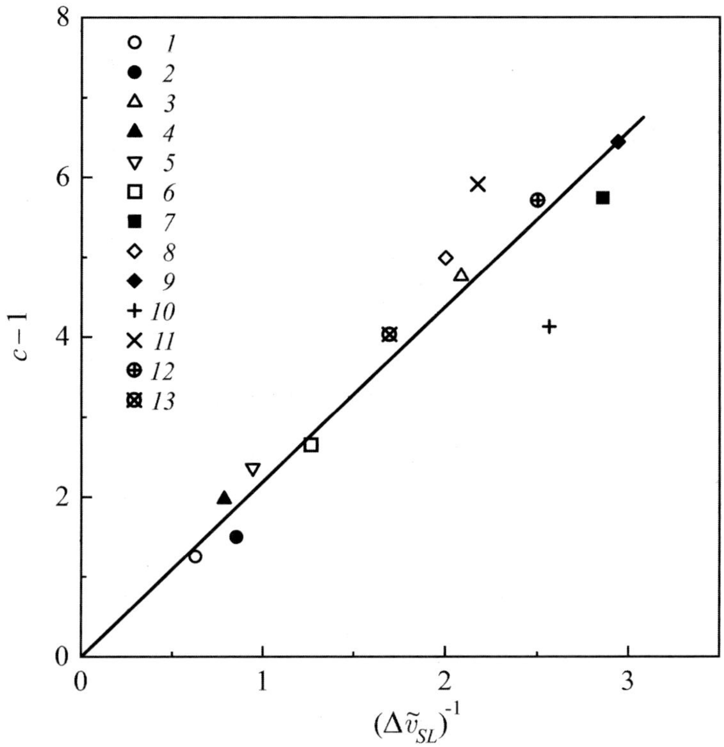

Figure 3.36: Correlation between the parameters $(c-1)$ and $(\Delta\widetilde{v}_{SL})^{-1}$ for alkali metal halides: (1) LiF, (2) LiCl, (3) NaF, (4) NaCl, (5) NaBr, (6) NaI, (7) KF, (8) KCl, (9) KBr, (10) RbF, (11) RbCl, (12) RbBr, (13) RbI.

The following comments can be made about the nature of the entropy term $(\Delta\widetilde{s}_1)_{SL}$. The difference between the liquid and the crystal does not consist only in the absence of a regularity in the mean positions of the molecules, i.e., in the existence of the long-range order of the crystal, but also in the existence of a large variety of "basic units" of short-range order in the liquids, and the resulting higher freedom in the coordination of the short-range neighbors. The latter difference is due to the fact that the liquid has no limitations with respect to the ordering of the molecules caused by the existence of a long-range ordering in a crystal. The development of such an approach to the interpretation of the structural properties of liquids has been initiated by the works of Bernal [200] in contrast to concepts employing ideas of a quasi-crystalline structure of the liquids.

More detailed information about the structural elements in a condensed system of many particles can be obtained by computer molecular-dynamic modeling [201]. The statistical analysis of the geometric characteristics of the so-called Voronoy polyhedra, constructed with account being taken of the positions of the neighboring atoms, shows significant differences in the structure of liquid and crystal already for such simple substances like argon [201, 202]. In addition, it turns out [203] that the transition from the liquid to the fcc crystal and back cannot be accomplished by just small shifts in the positions of the atoms: for such transitions a significant part of the atoms should be moved by a distance of about one atomic spacing. Crystallization is accompanied not only by the establishment of spatial regularity but simultaneously by a change of the short-range coordination of the atoms. Interpreting the liquid–crystal transition in such a way, it helps to understand the existence of deeply supercooled states of liquid and that nucleation is a necessary prerequisite for the crystallization to start. The parameters $\Delta\widetilde{s}_{SL} - \Delta\widetilde{s}_h$ and $(\Delta\widetilde{s}_1)_{SL}$ reflect the essential difference between liquid and crystal states.

Table 3.16: Parameters characterizing the melting process of alkali metal halides.

Substance	T_0, K	p_* MPa	c	Δv_{SL} cm^3/mol	$\Delta\widetilde{v}_{SL}$	$\Delta\widetilde{s}_{SL}$	Ref.
LiF	1121	4500	2.25	3.27	1.59	3.6	[207]
LiCl	878	1450	2.50	5.88	1.17	2.9	[208]
NaF	1265	1220	5.76	4.15	0.48	2.8	[209]
NaCl	1074	1500	2.97	7.55	1.27	3.8	[209]
NaBr	1014	1110	3.36	8.07	1.06	3.6	[209]
NaI	928	713	3.65	8.58	0.79	2.9	[209]
KF	1124	738	6.74	4.45	0.35	2.4	[210]
KCl	1043	598	5.99	7.20	0.50	3.0	[210]
KBr	1006	357	7.44	7.98	0.34	2.5	[210]
RbF	1071	1380	5.13	2.5	0.39	2.0	[211]
RbCl	991	560	6.91	6.72	0.46	3.2	[211]
RbBr	950	4400	6.71	7.26	0.40	2.7	[211]
RbI	913	560	5.03	8.05	0.59	3.0	[211]

Melting of one-atomic or similarly melting substances is characterized at $T = T_0$ by a value $(\Delta\widetilde{s}_1)_{SL} \cong 0.61$. For most of the substances consisting of molecules composed of several atoms the melting process is accompanied both by positional and orientational disordering, so additional degrees of freedom are reactivated. During this process both $\Delta\widetilde{s}_{SL}$ and $(\Delta\widetilde{s}_1)_{SL}$ are increased, but $c - 1$ and $(\Delta\widetilde{v}_{SL})^{-1}$ remain correlated. The correlation between these values is shown in Fig. 3.35. The mean-square deviation of experimental points from the linear dependence is of the order of 0.16. For a large group of substances $(\Delta\widetilde{s}_1)_{SL}$ is close to 2.60 (the angular coefficient of the straight line in Fig. 3.35), which is by two units larger than for positional melting. This excess can be attributed to the two additional vibration-rotational degrees of freedom of the molecular motion, if each degree of freedom has the melting entropy Δs_{SL} equal to R. The parameters c and p_* of the melting lines for the substances considered (cf. Fig. 3.35) are given in the Table 3.6. The main thermodynamic parameters, characterizing the melting of these substances at atmospheric pressure, are given in Table 3.15. This table also contains the values $\Delta\widetilde{s}_h$ and $\Delta\widetilde{s}_{SL} - \Delta\widetilde{s}_h$ for three substances calculated by using (p, v, T)-data found in the literature. The references to the works from where the respective (p, v, T)-data were taken are given in a separate column of the table. Similarly to the case of simple substances, the values $(\Delta\widetilde{s}_1)_{SL}$ and $\Delta\widetilde{s}_{SL} - \Delta\widetilde{s}_h$ are close to each other.

Equation (3.87) holds true also for ionic compounds [206]. A correlation between the quantity $(\Delta\widetilde{v}_{SL})^{-1}$ and $c - 1$ has been verified for alkali metal halides. The parameters

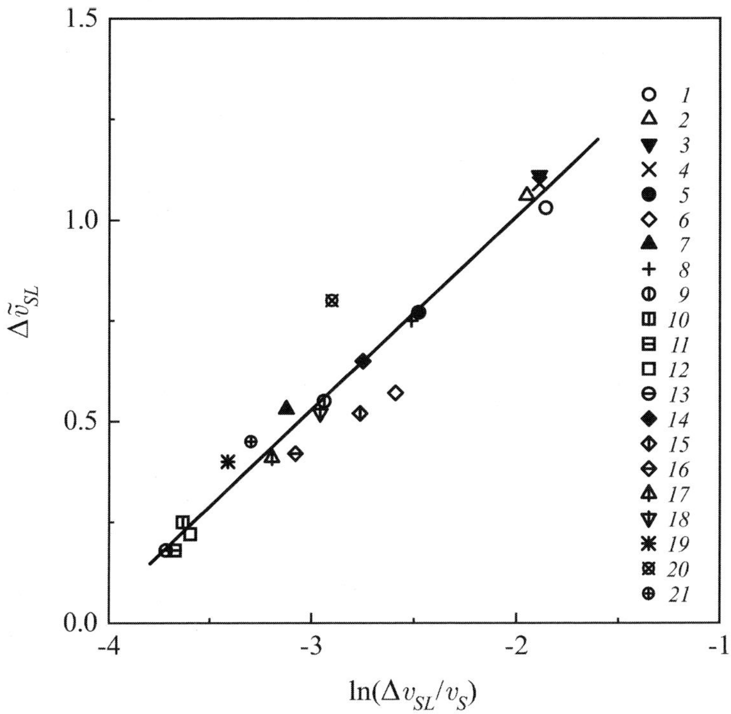

Figure 3.37: Correlation between the parameters $\Delta\widetilde{v}_{SL}$ and $\ln(\Delta v_{SL}/v_S)$ for simple substances. The different substances are as in Fig. 3.32.

of these compounds related to the melting process are given in Table 3.16. The table also contains the references to the literature from where the data for T_0, p_*, c, and Δv_{SL} were taken. Figure 3.36 shows for the studied ionic salts the correlation between the similarity parameter c and the inverse value of the reduced volume jump, $\Delta\widetilde{v}_{SL}$. The straight line in the figure corresponds to the correlation

$$c - 1 = 2.23(\Delta\widetilde{v}_{SL})^{-1} \ . \tag{3.99}$$

Another newly established correlation is connected with the volume jump [141, 193]. For simple substances the reduced volume jump $\Delta\widetilde{v}_{SL}$ at atmospheric pressure is correlated with the logarithm of the relative volume change during the melting process, $\Delta v_{SL}/v_S$. This correlation can be seen in Fig. 3.37. The dependence between the parameters $\Delta\widetilde{v}_{SL}$ and $\ln(\Delta v_{SL}/v_S)$ is close to a linear one and can be expressed by the following equation

$$\Delta\widetilde{v}_{SL} = 0.480 \ln\left(\frac{\Delta v_{SL}}{v_S}\right) + 1.955 \ . \tag{3.100}$$

The straight line in Fig. 3.37 corresponds to this equation. The mean-square deviation of the points from Eq. (3.100) is equal to 0.08.

The established correlations can be used for the determination of the melting curves of substances if the values of Δv_{SL} and $(\Delta v_{SL}/v_S)$ at normal pressure are known. For simple substances, we have the following method which allows us to calculate the parameters p_* and c of the unknown melting curve. Using the correlation Eq. (3.100) we first determine the value of the parameter $\Delta\widetilde{v}_{SL}$. Then, taking into account the definition Eq. (3.30) we obtain the parameter p_* via the relation

$$p_* = RT_0 \left(\frac{\Delta\widetilde{v}_{SL}}{\Delta v_{SL}} \right) . \tag{3.101}$$

Then, using the correlation between $(c-1)$ and $(\Delta\widetilde{v}_{SL})^{-1}$,

$$c - 1 = \frac{0.61}{(\Delta\widetilde{v}_{SL})^{-1}} \tag{3.102}$$

and the known value of $\Delta\widetilde{v}_{SL}$, we determine the value of the parameter c, i.e., the whole melting line in the approximation of the Simon equation. The knowledge of the quantities $\Delta v_{SL}(T = T_0)$, p_*, and c allows us to calculate the values of melting entropy and enthalpy at $T = T_0$.

For polyatomic substances, an estimate of the parameters c and p_* of the unknown melting curve can be made by using Eq. (3.49). In order to estimate p_*, one needs to know the critical pressure, p_c, on the liquid–vapor equilibrium coexistence curve. The value of the parameter c is estimated using the correlation established by us.

$$c - 1 = \frac{2.60}{(\Delta\widetilde{v}_{SL})^{-1}} . \tag{3.103}$$

The same method of computation of the parameters c and p_* can be employed also for ionic compounds. The critical pressure p_c of these substances was estimated using the rectilinear diameter rule [206].

Table 3.17: Estimates of the parameters p_* and c of the melting curves for several substances.

Substance	T_0, K	$p_{*(calc)}$, MPa	$p_{*(exp)}$, MPa	$c_{(calc)}$	$c_{(exp)}$	Ref.
Fluorine	53.48	216	250	2.37	2.18	[147]
Copper	1357.7	15400	13400	2.30	2.44	[137]
Silver	1235.1	9590	9200	2.16	2.07	[137]
Chlorine	172.2	416	494	2.46	2.11	[148]
Monofluorobenzene	231.0	448	601	2.09	1.79	[67]
Hexafluorobenzene	278.1	495	454	1.93	1.85	[67]
NaCl	1074	1400	1500	2.76	2.97	[209]
KBr	1006	441	357	7.56	7.44	[210]

Table 3.17 contains the estimates of the parameters p_* and c for simple substances, polyatomic compounds and ionic salts obtained using the described methods. For comparison, the values of these parameters are given that were obtained from experimental data by the

dependencies $p_{SL} = f(T)$ in a wide temperature range. The references to original papers dealing with the melting lines are given in a separate column of the table. The comparison of the values of the parameters p_* and c, obtained from experimental data and the ones predicted using the established correlations, shows an acceptable agreement between the results of the computations and experiment. Note that in order to estimate p_* for molecular (nonionic and nonmetallic) substances one can also use the correlation Eq. (3.44) between p_* and the internal pressure $p_i(1)$ in the liquid at $\widetilde{T} = 1$. For the initial slope of the melting curve in the approximation of the Simon equation ($(dp/dT_{SL}) = p_* c/T_0$), the deviation of calculated and experimental data is of the order of 3% for inert gases and alkali metals and of the order of $5 - 8\%$ for the other simple substances shown in Table 3.14. In the case of molecular compounds (Table 3.15), the deviation of calculated values of the quantity (dp/dT_{SL}) from the experimental data is of the order of $10 - 15\%$. For the salts, this deviation varies from one to several tens per cent. Taking into account that the correlation coefficients for the established relations, Eqs. (3.49), (3.99)–(3.103), are near to one, we can suppose that the observed deviations are mostly caused by the insufficient precision in the determination of the parameters p_* and c based on the experimental data. For the salts, a significant contribution to such deviation can be introduced by the method of estimation of the critical pressure p_c.

The established correlations reveal the important role of the volume as a decisive quantity for the equilibrium melting curve and the advantage of using the shifted pressure scale determined via Eq. (3.19), where the value p_*, which has the meaning of a characteristic internal pressure, effectively takes into account the interatomic (intermolecular) attraction forces in the condensed phase and defines the scale of these forces. The shifting of the origin of the pressure coordinate into the pole, $p_* = -p_{SL}(T \to 0)$, $p' = p + p_*$, is the basis for the low-temperature asymptotics of the melting lines given by

$$\frac{dp}{dT_{SL}} \to 0 \qquad \text{at} \qquad T \to 0\,, \quad p' \to 0\,, \tag{3.104}$$

and also for the thermodynamic similarity in the variables (p'_1/p'_2), (T_1/T_2) or $\widetilde{p}_{SL}$ and $\widetilde{T}_{SL}$. The parameter c in Eq. (3.3) defines a group of substances being similar with respect to the melting process. It is defined by Eqs. (3.24) or (3.31) containing the characteristic pressure p_*. The values of p_* in the specification of the correlations were determined using the Simon equation, so all the quantitative results are obtained in this approximation.

Summarizing the results of the thermodynamic analysis of the crystal–liquid phase equilibrium curve, we would like to note some key points. Anchoring this line to the asymptotic low-temperature limit $T \to 0$, $p \to -p_* < 0$ allowed us to introduce a shifted pressure scale $p' = p + p_*$, where $p_* > 0$ is an individual parameter specific for the substance under consideration. It is defined by an extrapolation of the dependence $p_{SL} = f(T)$ beyond the triple point into the region of metastable (stretched) states of both phases. The value of the parameter p_* is used as a pressure scale for the transition to the dimensionless variables $\widetilde{p}$, $\widetilde{v}$, and $\widetilde{s}$ and allows us to arrive at the correlations expressing the thermodynamic similarity of substances in the melting process. For $T \to 0$, the condition of phase equilibrium, $\mu_S(p,T) = \mu_L(p,T)$, goes over into the "mechanical" identity $\Delta u = p_* \Delta v$, which does not contain the entropy. In this limit, the pressure p_* coincides with the internal pressure p_i of the condensed phase, i.e., the relations

$$p_i = \left(\frac{\partial u}{\partial v}\right)_{T=0} = p_* = \left(\frac{\Delta u}{\Delta v}\right)_{T=0} \tag{3.105}$$

hold. Reference points for the melting curves of different substances are the points with the same values of the ratios $\widetilde{p} = p'/p_*$ and $\widetilde{T} = T/T_0$. For $\widetilde{T} = 1$, the universality of splitting of the entropy into two parts has been established. In this case, we get $\Delta\widetilde{s}_{SL} = \Delta\widetilde{v}_{SL} + (\Delta\widetilde{s}_1)_{SL}$. Using the Simon equation Eq. (3.3) and the approximation for $\Delta v(T)$, one can find the variation of the "universal" component $(\Delta\widetilde{s}_1)_{SL}$ with temperature.

The Simon equation qualitatively correctly represents the line $p_{SL} = f(T)$ for normally melting substances. The parameter p_* is determined by experimental data for $p > 0$ by the condition to reach the best agreement of Eq. (3.3) with the experiment. The pole, $T = 0$, $p' = 0$, is located near to the boundary of stability (spinodal) of the stretched crystal and liquid. Here one may see an analogy with the liquid–vapor critical point. The entropy differences of both phases also vanish, $\Delta s_{SL} \to 0$; however, the volume differences remain finite, i.e., $\Delta v_{SL} \neq 0$.

3.11 Melting and Crystallization of Small Particles

3.11.1 Thermodynamic Aspects

The specific features of the phase transitions in small particles are due to the fact that in these particles a noticeable part of the molecules is located in the surface layer. This property introduces considerable changes into the energetics of the system. The known phenomenological models, describing the dimensional effect at the melting of the particles, are based on various approximations [212–216]. The results of the different theoretical approaches significantly depend on the choice of the melting criterion. Two different cases should be distinguished here:

1. a small crystal of radius r is located in a large mass of the melt;
2. small droplets and crystals of similar sizes, R, are surrounded by their own vapor.

We will consider, now, the conditions of thermodynamic equilibrium for both cases assuming the crystal to be of spherical shape.

Analysis of the First Case: At a given external pressure, $p_A = p_0$, the crystal and the liquid phases are in equilibrium at a planar interface at the temperature $T_A = T_0$ (see Fig. 3.38) and the chemical potentials are equal

$$\mu_S(T_0, p_0) = \mu_L(T_0, p_0) \ . \tag{3.106}$$

For a small crystal of radius r at an unstable equilibrium with the melt the following identities hold true

$$\mu_S(T_B, p_S) = \mu_L(T_B, p_A) \ , \tag{3.107}$$

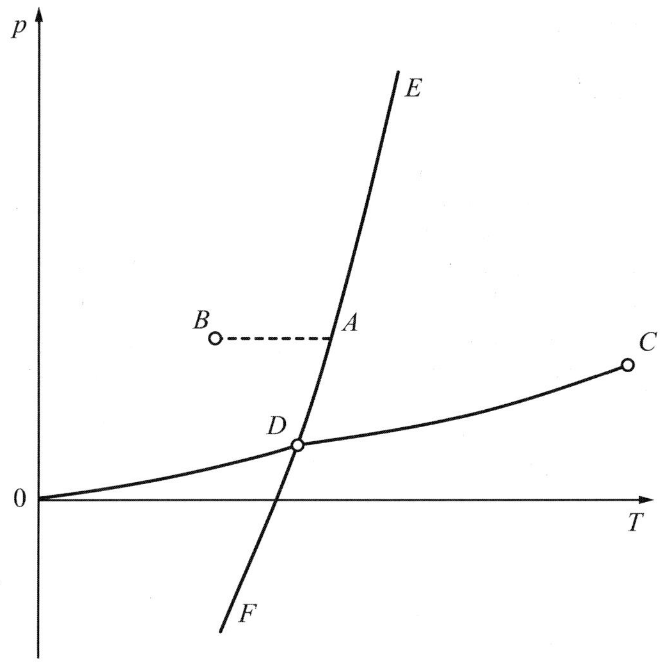

Figure 3.38: Phase diagram for the equilibrium of a small crystalline particle in the supercooled melt (point B).

$$p_S - p_A = \frac{2\sigma_{SL}}{r} \, . \tag{3.108}$$

The point B in Fig. 3.38 refers to the state of the supercooled liquid. Subtracting Eq. (3.106) from Eq. (3.107) and expanding the differences of chemical potentials as series in temperature and pressure retaining only the first terms, we obtain

$$-s_S(T_B - T_0) + v_S(p_S - p_0) = -s_L(T_B - T_0), \tag{3.109}$$

or

$$\Delta h_{SL}\frac{(T_B - T_0)}{T_0} + \frac{2\sigma_{SL}v_S}{r} = 0 \, , \tag{3.110}$$

where $\Delta h_{SL} = (s_L - s_S)T_0$ is the melting heat at the given pressure, $p_A = p_0$. Introducing further the notation $\Delta T = T_0 - T$, we have the following expression for the decrease of the equilibrium temperature of the melt with a crystal of radius r

$$\Delta T = \frac{2\sigma_{SL}v_S T_0}{\Delta h_{SL} r} \, . \tag{3.111}$$

If we consider in such way the thermodynamic equilibrium of a droplet of the melt inside a large volume of a crystal phase, i.e., when internal and external phases are exchanged, then the sign of ΔT will change and we get

$$\Delta T = T_0 - T = -\frac{2\sigma_{SL} v_S T}{\Delta h_{SL} r} \; . \tag{3.112}$$

On the (T, p)-phase diagram, the regions of nucleation for the liquid and for the crystal are located at different sides of the binodal (EF) shown in Fig. 3.38. If the condensed system is a small thermodynamic object, then both phase transitions (crystallization and melting) proceed at a supercooling $\Delta T = T_0 - T > 0$. This topic is discussed in more detail below.

Analysis of the Second Case: Let us consider, now, small droplets (liquid and crystal) of the same mass M, surrounded by vapor. The temperature is supposed to be below the triple point (point D in Fig. 3.38), where the pressure of the vapor, $p_V < p_D$, is small. A weak interaction of the droplets takes place through the vapor. The pressure in the droplets is higher than in the vapor. We have

$$p_S - p_V = \frac{2\sigma_{SV}}{R_S} \; , \qquad p_L - p_V = \frac{2\sigma_{LV}}{R_L} \; . \tag{3.113}$$

For small values of R one can replace in Eq. (3.113) the quantity p_V by the pressure in the triple point p_D without affecting the result considerably.

For the difference between the chemical potential of the crystalline particle and the chemical potential of a large crystal at temperature T_D we have

$$\begin{aligned} \mu_S(T, p_S) - \mu_{S0}(T_D, p_D) &\cong \left(\frac{\partial \mu_S}{\partial T}\right)(T - T_D) + \left(\frac{\partial \mu_S}{\partial p}\right)(p_S - p_D) \\ &= -s_S(T - T_D) + v_S(p_S - p_D) \; . \end{aligned} \tag{3.114}$$

Similarly, for a liquid droplet, we arrive at

$$\mu_L(T, p_L) - \mu_{L0}(T_D, p_D) \cong -s_L(T - T_D) + v_L(p_L - p_D) \; . \tag{3.115}$$

At phase equilibrium in the triple point the following equality holds

$$\mu_{S0}(T_D, p_D) = \mu_{L0}(T_D, p_D) = \mu_{V0}(T_D, p_D) \; . \tag{3.116}$$

Therefore, in order to get an equilibrium of the droplets the following condition should be satisfied

$$-s_D(T - T_D) + v_S(p_S - p_D) = -s_L(T - T_D) + v_L(p_L - p_D) \tag{3.117}$$

or, taking into account Eqs. (3.109) and (3.113),

$$\Delta h_{SL} \frac{T_D - T}{T_D} = \frac{2\sigma_{SV} v_S}{R_S} - \frac{2\sigma_{LV} v_L}{R_L} = \frac{2 v_S}{R_S}\left[\sigma_{SV} - \sigma_{LV}\left(\frac{v_L}{v_S}\right)^{2/3}\right] \; . \tag{3.118}$$

If we neglect the difference between the specific volumes of liquid, v_L, and crystal, v_S, and, hence, between the radii R_L and R_S at a given mass of the droplet, and take into account the relation $\sigma_{SL} = \sigma_{SV} - \sigma_{LV}$, we obtain the well-known Thomson formula (which coincides formally with Eq. (3.111))

$$\frac{\Delta T}{T_D} = 1 - \frac{T}{T_D} = \frac{2\sigma_{SL} v}{\Delta h_{SL} R} . \tag{3.119}$$

The temperature T represents the boundary between the regions of thermodynamically preferable existence of liquid and crystal states of a droplet of radius R. Both states represent supercooled states with respect to the triple point. Under the assumptions made, Eq. (3.119) holds true also for any arbitrary pressure p in the range $0 < p < p_D$ along the respective segment of the equilibrium melting line.

Some analogue of the Thomson formula with a different numerical coefficient on the right-hand side of Eq. (3.119) can be obtained without the application of the condition of equilibrium with the vapor phase. For such purposes, we write down the difference, $\Delta\Phi_{S/L}$, of the Gibbs thermodynamic potential of a crystal particle as compared with a liquid droplet of the same radius (it is assumed that $v_S = v_L$ holds), separating the volume and surface parts in $\Delta\Phi_{S/L}$ as

$$\Delta\Phi_{S/L} = -\frac{4}{3}\pi R^3 \rho \Delta\varphi + 4\pi R^2 \sigma_{SV} - 4\pi R^2 \sigma_{LV} , \tag{3.120}$$

where

$$\Delta\varphi = \Delta h_{SL} \frac{\Delta T}{T_0} , \qquad \rho = \frac{1}{v} \tag{3.121}$$

holds. For a unit of mass we have

$$\frac{\Delta\Phi_{S/L}}{M} = -\Delta h_{SL} \frac{\Delta T}{T_0} + \frac{3\sigma_{SL} v}{R} \qquad \text{with} \qquad M = \frac{4\pi}{3} R^3 \rho \tag{3.122}$$

and as before it is supposed that $\sigma_{SL} = \sigma_{SV} - \sigma_{LV}$ is valid. As follows from Eq. (3.122), the crystalline particle is thermodynamically preferable as compared with the liquid drop

$$\Delta\Phi_{S/L} < 0 \qquad \text{for} \qquad \Delta T < \frac{3\sigma_{SL} T_0 v}{\Delta h_{SL} R} . \tag{3.123}$$

The indifferent case (crystal and drop are thermodynamically equivalent states) is described by the relation

$$\frac{\Delta T}{T_0} = \frac{3\sigma_{SL} v}{\Delta h_{SL} R} . \tag{3.124}$$

Instead of a factor 2 in the Thomson formula Eq. (3.119), here we obtained a similar expression but with a numerical coefficient 3. This difference reflects the approximate character (the ambiguity) of the thermodynamic approach to the analysis of the problem of phase transitions in small particles.

3.11.2 Kinetic Aspects

Now we complete the thermodynamic estimates with kinetic considerations concerning phase transitions in small particles. The events of melting/crystallization of separate particles at a low rate of sublimation near the melting temperature can be considered as independent due to the weak interaction between them (mostly through the thermostat). The Thomson formula Eq. (3.119) gives only an approximate estimate of the decrease of the melting temperature of nanosized particles. For the given size, R, and temperature, $T < T_0$, the quasi-stationary phase state of the ensemble of particles is defined by the probabilities of forward and backward transitions between melting and crystallization of the particles (melting $\leftrightarrow$ crystallization). The thermodynamic system under consideration is a separate drop located in a thermostat. The creation of only one viable nucleus of the competing phase is sufficient to generate the phase transition inside the particular drop. There is a correlation between the mean waiting time of occurence of such a nucleus, $\bar{\tau}$, and the nucleation rate, J_1 (in $\mathrm{s}^{-1}\mathrm{cm}^{-3}$), given via

$$\frac{1}{\bar{\tau}} = V J_1 \,, \tag{3.125}$$

where V is the volume of the droplet.

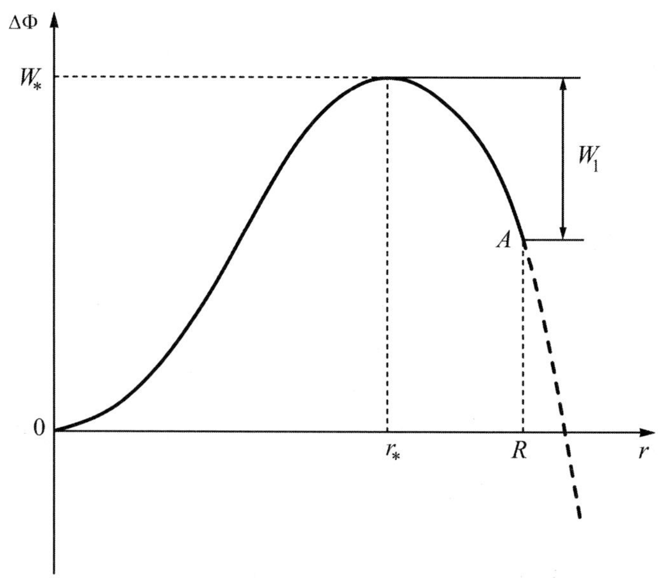

Figure 3.39: Dependence of the Gibbs free energy barrier, $\Delta\Phi$, on the radius of a crystalline particle at a temperature $T < T_0$.

Due to the effect of the surface tension, at $T < T_0$ both phases (crystal and liquid) are metastable with respect to the conditions of a phase equilibrium of a large mass at planar interfaces. Both forward and backward transitions in a nanosized particle are connected with an overcoming of the barrier of the Gibbs free energy, $\Delta\Phi$. The height of the barrier, W_*, depends on temperature (the external pressure can be considered as constant and close to zero). At $T < T_0$, the dependence of $\Delta\Phi$ on the radius of the crystal nucleus, r, has a form as shown in Fig. 3.39. The dashed extension of the curve, $\Delta\Phi(r)$, corresponds to the crystal in an "infinitely" extended liquid. The maximum, $\Delta\Phi = W_*$, corresponds to the critical nucleus, $r = r_*$. But for the small particle in which the transition proceeds, the size of the system is limited by the value $r = R$ (the radius of the droplet). In this case (for the given supercooling), the height of the barrier for the backward transition W_1 can be less than W_* as is shown in Fig. 3.39. By this reason, a situation may occur when both transitions are competing kinetically and the result depends on the rate of these transitions initiated by thermal fluctuations. The nucleation mechanism of crystal and liquid is also different in small particles. The crystal nucleus can emerge at any place of the liquid volume, while liquid islands are formed preferentially on the surface of the crystal. Such single liquid lenses can combine into a continuous layer with a width, δ. If we denote the "critical" depth of this layer by δ_*, where $\delta_* = R - r_*$ holds, and write the expression for W_1 in a thermodynamical approximation taking into account volume and surface contributions to the Gibbs free energy, then instead of the Thomson formula Eq. (3.119) we obtain the following approximation [6, 217]

$$\Delta T = T_0 - T = \frac{2\sigma_{SL} T_0 v_S}{\Delta h_{SL}(R - \delta_*)} , \tag{3.126}$$

where

$$\delta_* \approx \left(\frac{3W_1}{16\pi\sigma_{SL}} \right)^{1/2} \tag{3.127}$$

holds. The value of δ_* can be estimated by using the method of calculation of the stationary nucleation rate via the homogeneous nucleation mechanism.

Here we restrict the analysis only to the surface fluctuations in the course of melting of crystalline particles of radius R. The fluctuation-initiated formation of a liquid layer of critical depth, δ_*, on the surface of the particle requires overcoming of the free energy barrier W_1. Let us denote by i the number of atoms in a surface layer of the particle, where the relevant fluctuations take place. Then, using the traditional approach of the theory of homogeneous nucleation [7, 8] one can obtain [6, 218] expressions for the nucleation rate J_1 on the surface of the solid particle of volume V, or the mean waiting time of melting of the crystalline particle, $\bar{\tau} = (VJ_1)^{-1}$

$$VJ_1 = Zi \left(\frac{k_B T}{h} \right) \exp\left(-\frac{E}{k_B T} \right) \exp\left(-\frac{W_1}{k_B T} \right) , \tag{3.128}$$

where h is the Planck constant and E is the energy required to activate the process of transfer of molecules from the crystal phase to the liquid one.

For a nanosized particle, the full number of atoms, N, exceeds the number of "surface" atoms by not more than one or two orders of magnitude. The nonequilibrium factor, Z, retains its usual value $Z \cong 10^{-2} - 10^{-3}$ s. Assuming that the characteristic time of melting of the particle is about a second, we obtain from Eq. (3.128) the estimate $W_1 \cong 30 k_B T$ for the energetic barrier. Using this value in Eq. (3.127), we obtain for a number of metals an estimate of the critical depth of the liquid layer, δ_*, on the crystal nucleus of the order of 0.5–0.6 nm.

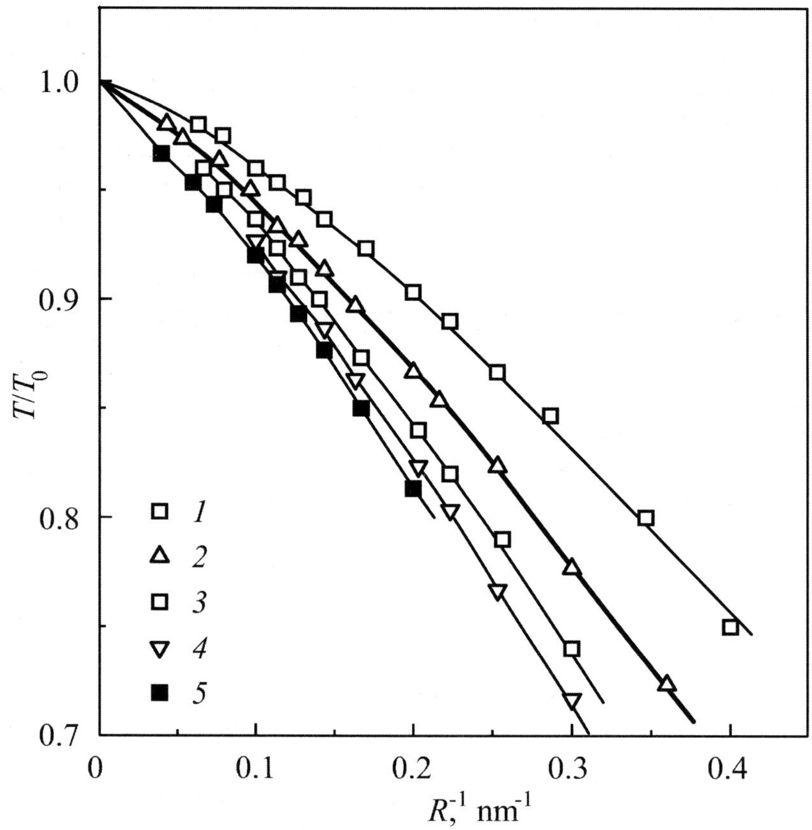

Figure 3.40: Dependence of the relative melting temperature of metallic particles on the curvature of the surface: (1) lead, (2) tin, (3) indium, (4) bismuth, (5) (β)-gallium.

The random, fluctuating character of the melting process of small particles requires a statistical treatment of the experimental data, i.e., multiple observations of melting of one particle in reproducible conditions, or the observation of the evolution of an ensemble of independent particles of the same size. The second approach is more effective for metallic island films evaporated on a substrate. Electronographic methods can be used to identify the phase state of the particles. For the first time, this method was used by Takagi in 1954 [219]. For films of lead, tin and bismuth of about 5 nm thickness, a decrease of the melting temperature was discovered by 31, 30 and 23 K, respectively. In Ref. [220], the melting temperature of tin and bismuth in thin layers was determined.

Below we summarize results obtained at the Institute of Thermal Physics of the Ural Branch of the Russian Academy of Sciences. References to the original publications can be found in Ref. [6]. The analysis of melting and crystallization of small metallic particles (4–100 nm in diameter) has been carried out using the scanning registration of the integral intensity of electron scattering in dependence of the change in temperature of the island films [217].

The samples were obtained by evaporation and condensation of the metal onto an amorphous carbon substrate in the camera of the electronograph. After evaporation, the samples were heated to temperatures slightly above their melting temperature. This method allowed us to obtain a shape of the islands close to spheres. The intensity of the chosen reflex of the diffraction picture has been observed in the cycles of heating/cooling of the sample. The melting of the island films was accompanied by the disappearance of the diffraction reflex of the crystal state. In the course of the crystallization of the sample, in contrast, its appearance was observed. In the approximation of a kinematic electron scattering, the integral intensity, I_{exp}, of the diffraction reflex is proportional to the mass fraction of the crystalline substance in the film [221]

$$\frac{M_S(T)}{M_0} = \frac{I_{exp}(T)}{I(T)} , \tag{3.129}$$

where $M_S(T)$ is the mass of the crystallized particles scoped by the electron beam, M_0 is the total mass of particles (both liquid and crystal) scoped by the beam, and $I(T)$ corresponds to the intensity of a completely crystalline sample. The nucleation rate for the crystallization of droplets during the cooling with the rate of $\dot{T}$ has been calculated based on experimental data for the ratio $M_S(T)/M_0$ using the relation

$$\frac{M_S(T)}{M_0} = 1 - \frac{4\pi}{\bar{V}} \int_0^\infty \exp\left[-\frac{4\pi J R^3}{3\dot{T}(d\ln J/dT)}\right] f(R) R^3 dR . \tag{3.130}$$

The size distribution of droplets, $f(R)$, and the mean volume of particles, $\bar{V}$, in the island films have been determined based on photographs of the sample obtained with an electron microscope. The data for the dependence of the nucleation rate on temperature, $J(T)$, in supercooled tin, which were obtained in the experiments with island films, are shown in Fig. 3.28. There also a comparison with the experimental data is given obtained for samples of submicrometer size and with the results of calculations employing the homogeneous nucleation theory. As follows from the figure, the experimental data obtained by different methods for the samples of different sizes are well described by the theory of homogeneous nucleation assuming a constant value of the surface tension, σ_{SL}. This result is an indication of the homogeneous mechanism of the crystallization of supercooled tin droplets.

The electronographic method of studying the processes of melting and crystallization of small metal particles allows us to establish the dependence of phase transition temperatures on the size of the particles. Figure 3.40 shows for a group of metals the dependence of the relative melting temperature, T/T_0, on the curvature of the particle surface. The experimental dependence of the temperature on the value of R^{-1} is different from the linear dependence given by the Thomson formula Eq. (3.119). The solid lines in Fig. 3.40 correspond to the following correlation [222]

$$T = T_0 - \frac{2a\sigma_{SL}T_0 v_S}{\Delta h_{SL}(R - b\delta_*)} , \tag{3.131}$$

which is based on the assumption of the stochastic character of the melting of small crystal particles. In contrast to Eq. (3.126), Eq. (3.131) contains the empiric constants a and b, which

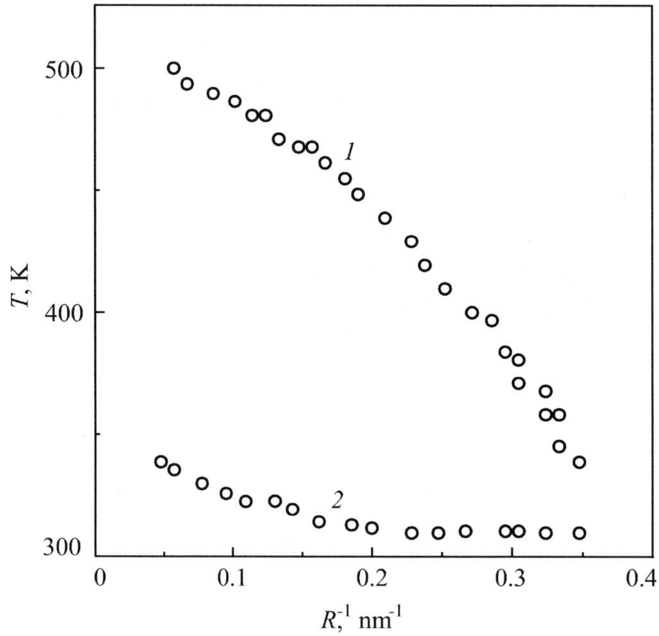

Figure 3.41: Dependence of the temperatures of melting (1) and crystallization (2) on curvature of the surface of small tin particles ($T_0 = 505$ K).

are determined from the processing of experimental data for the dependence of the melting temperature on the curvature of the surface of the particles and characterize the deviation of the shape of the particles from spherical.

Figure 3.41 shows the dependencies of the temperatures of melting and crystallization on the curvature of the surface of small particles, which were obtained in experiments with tin island films at a cooling rate of 0.1 K/s. The point of intersection of the two curves with the increase of curvature, R^{-1}, corresponds to a disappearance of the temperature hysteresis during melting and crystallization of the particles. This property is a consequence of the equality of the energetic barriers of melting and crystallization or, in other words, the equiprobable character of both phase transitions for a small particle.

4 Phase Transitions in Solutions

4.1 Generalized Clausius–Clapeyron Equation for Solutions

The Clausius–Clapeyron equation Eq. (1.2) can be generalized for the case when the thermodynamic system is not a single-component one and/or when it is subjected not only to the external pressure, but also to forces of other nature (e.g., electric or magnetic field intensity). Let us consider the form of the Clausius–Clapeyron equation for a two-phase, two-component system [223], i.e., for a solution containing molecules of two kinds. Here N_1 and N_2 are the numbers of moles of the different components in the system, and $x_i = N_i/(N_1 + N_2)$ is the molar fraction of the respective component. One and two primes specify the values related respectively to the first and second phase. For each phase we have $x_1 + x_2 = 1$. Phase equilibrium presumes equality of the chemical potentials of the independent components in the coexisting phases, i.e.,

$$\mu_1'(T, p, x_1') = \mu_1''(T, p, x_1'') , \tag{4.1}$$

$$\mu_2'(T, p, x_2') = \mu_2''(T, p, x_2'') . \tag{4.2}$$

The equilibrium is retained for small variations of the thermodynamic variables, so we have

$$d\mu_1' = d\mu_1'' , \qquad d\mu_2' = d\mu_2'' . \tag{4.3}$$

The change of the chemical potentials as expressed by Eq. (4.3) is determined, in general, by the change of temperature, pressure and concentration in the coexisting phases. We may write

$$d\mu_i = \left(\frac{\partial \mu_i}{\partial T}\right)_{p,x_i} dT + \left(\frac{\partial \mu_i}{\partial p}\right)_{T,x_i} dp + \left(\frac{\partial \mu_i}{\partial x_i}\right)_{T,p} dx_i , \qquad i = 1, 2 . \tag{4.4}$$

One can substitute into this expression the values of the derivatives $(\partial \mu_i/\partial T)$ and $(\partial \mu_i/\partial p)$ using the equality of cross derivatives in Gibbs's fundamental equation

$$dG = -SdT + Vdp + \mu_1 dN_1 + \mu_2 dN_2 \tag{4.5}$$

for each phase. Then Eqs. (4.3) take the form

$$-(\bar{s}_1'' - \bar{s}_1')dT + (\bar{v}_1'' - \bar{v}_1')dp = \frac{\partial \mu_1'}{\partial x_1'} dx_1' - \frac{\partial \mu_1''}{\partial x_1''} dx_1'' \tag{4.6}$$

Crystal–Liquid–Gas Phase Transitions and Thermodynamic Similarity. Vladimir P. Skripov and Marsz Z. Faizullin

ISBN: 3-527-40576-3

and

$$-(\bar{s}_2'' - \bar{s}_2')dT + (\bar{v}_2'' - \bar{v}_2')dp = \frac{\partial \mu_2'}{\partial x_1'}dx_1' - \frac{\partial \mu_2''}{\partial x_1''}dx_1'' \,, \tag{4.7}$$

where $\bar{s}_i = (\partial S/\partial N_i)_{T,p}$, $\bar{v}_i = (\partial V/\partial N_i)_T$ are respectively the partial molar entropy and volume of the component i in the two phases specified in Eqs. (4.6) and (4.7) by one or two primes. The molar fraction of the first component was taken as the independent concentration in Eqs. (4.6) and (4.7).

Following Ref. [224], we multiply the first equation of the system of equations Eqs. (4.6) and (4.7) by x_1'', the second by x_2'' and then take the sum. The right-hand side of the obtained sum is then simplified to

$$x_1''(d\mu_1'')_{T,p} + x_2''(d\mu_2'')_{T,p} = 0 \,. \tag{4.8}$$

After these modifications we have

$$\begin{aligned} &- [x_1''(\bar{s}_1'' - \bar{s}_1') + x_2''(\bar{s}_2'' - \bar{s}_2')]dT + [x_1''(\bar{v}_1'' - \bar{v}_1') + x_2''(\bar{v}_2'' - \bar{v}_2')]dp \\ &\qquad = x_1''\frac{\partial \mu_1'}{dx_1'}dx_1' + x_2''\frac{\partial \mu_2'}{\partial x_1'}dx_1' \,. \end{aligned} \tag{4.9}$$

We denote the phase with two primes as the high-temperature phase, e.g., the vapor being in equilibrium with the liquid, or liquid being in equilibrium with the crystal. The value

$$L = T[x_1''(\bar{s}_1'' - \bar{s}_1') + x_2''(\bar{s}_2'' - \bar{s}_2')] \tag{4.10}$$

has the meaning of the heat of the phase transition per mole of the solution, which has the composition of the high-temperature phase (with accuracy of the contribution of the heat of mixing). Accordingly the jump of the volume at the phase transition is defined by the following expression

$$\Delta v = x_1''(\bar{v}_1'' - \bar{v}_1') + x_2''(\bar{v}_2'' - \bar{v}_2') \,. \tag{4.11}$$

Substituting Eqs. (4.10) and (4.11) into Eq. (4.9), one obtains the Clausius–Clapeyron equation for a binary system as

$$-\frac{L}{T}dT + \Delta v dp = x_1''\frac{\partial \mu_1'}{\partial x_1'}dx_1' + x_2''\frac{\partial \mu_2'}{\partial x_1'}dx_1' \,. \tag{4.12}$$

This relation has been obtained in a somewhat different form by van der Waals [225] and is known as the van der Waals equation for a two-phase solution. If x_1' is kept constant, then Eq. (4.12) takes the form of the ordinary Clausius–Clapeyron equation $(\partial p/\partial T)_{x_1'} = L/T\Delta v$. If the values $(\partial p/\partial T)_{x_1'}$ and Δv are determined experimentally, then one can determine the heat of the phase transition, L, for the solution.

It can be seen from the above considerations that in order to change the independent variable x_1' to x_1'' one needs to use the Gibbs–Duhem equation Eq. (4.8) for the low-temperature phase. The phase transition heat and the volume jump are determined in this case by the composition of the phase specified by one prime.

For the practical use of Eq. (4.12) one has to have at one's disposal a model of the solution, which allows one to find the derivatives $(\partial\mu_1'/\partial x_1')_{T,p}$ and $(\partial\mu_2'/\partial x_1')_{T,p}$ as explicit functions of T, p, and x_1'. In the model of an ideal solution [226, 227] the dependence of the chemical potential on concentration is represented by a separate term like

$$\mu_i = \mu_{0i}(T,p) + RT\ln x_i \ , \tag{4.13}$$

so we have $(\partial\mu_1'/\partial x_1') = RT/x_1'$ and $(\partial\mu_2'/\partial x_1') = -RT/x_2'$. Introducing the notations

$$b_p = \frac{L}{RT^2}\frac{\partial T}{\partial x_1'} \ , \qquad b_T = \frac{\Delta v}{RT}\frac{\partial p}{\partial x_1'} \ , \tag{4.14}$$

we obtain for Eq. (4.12)

$$-b_p + b_T = \frac{x_1''}{x_1'} - \frac{x_2''}{x_2'} \ . \tag{4.15}$$

This equation can be solved in terms of the coefficient of substance distribution between the phases coexisting in equilibrium, $\beta = (x_1''/x_1')$, or in terms of the separation coefficient

$$\alpha = \frac{x_1''/x_1'}{x_2''/x_2'} \ . \tag{4.16}$$

For $p =$ constant, we have

$$\beta = 1 - b_p(1 - x_1') \ , \qquad \alpha = \frac{1 - b_p(1 - x_1')}{(1 + b_p x_1')} \ . \tag{4.17}$$

If the temperature is constant, one needs to substitute into Eq. (4.17) the value $-b_T$ from Eq. (4.14) instead of b_p. The phase transition heat is considered as additive, $L = L_1x_1'' + L_2x_2''$, as by the definition of an ideal solution the mixing heat h^M is equal to zero. Having at one's disposal Eqs. (4.17), by the known phase equilibrium line for the low-temperature phase $x_1' = x_1'(T; p = \text{constant})$ or $x_1' = x_1'(p; T = \text{constant})$ one can plot the second line of the phase diagram for an ideal solution: $x_1'' = x_1'\beta$. This plot is made by single points $\{x_1', T, p\}$ using the thermodynamic factor b_p or b_T related to these points.

If we take as a basic line the curve of the phase diagram, which is related to the high-temperature phase, then for the distribution and separation coefficients at $p =$ constant we obtain the following expressions

$$\beta = \frac{x''}{x'} = \frac{1}{1 + b_p(1 - x_1'')} \ , \qquad \alpha = \frac{1 - b_p x_1''}{1 + b_p(1 - x_1'')} \ , \tag{4.18}$$

where the relations $b_p = (L/RT^2)(\partial T/\partial x_1'')$ and $L = L_1x_1' + L_2x_2'$ hold. The transition to the case $T =$ constant is accomplished by changing b_p to $-b_T$.

In practice, phase diagrams plotted at constant pressure of the system or constant temperature are usually employed (see Fig. 4.1). In the approximation of ideal solutions both cases are described by correlations like Eqs. (4.17) and (4.18). The difference is in the factor b_i, $i = p, T$. For $T =$ constant this factor contains the volume change Δv at the

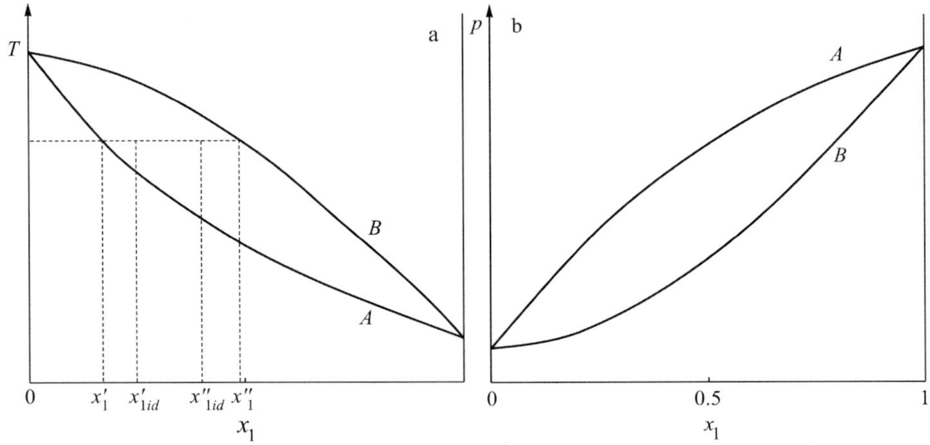

Figure 4.1: Liquid (A) - vapor (B) phase diagram at (a) constant pressure (p = constant) and (b) constant temperature (T = constant). x_1 denotes the molar fraction of the first component ($x_1 + x_2 = 1$).

phase transition, expressed through the partial molar volumes of the components Eq. (4.11): $b_T = (\Delta v/RT)(\partial p/\partial x'_1)$. Assuming that the solution is ideal, we substitute into Eq. (4.11) the molar volumes of vapor, v''_i, and liquid, v'_i, on the binodal curves of the components, $i = 1, 2$, for the given pressure. For the liquid–vapor system the calculation of Δv is simplified at temperatures significantly below the critical point of the low-boiling component. In this case neglecting the molar volume of the liquid phase v' in comparison with the molar volume of the vapor phase we have

$$\Delta v = v'' - v' \cong v'' = \frac{RT}{p}, \qquad b_T = \frac{1}{p}\left(\frac{\partial p}{\partial x'_1}\right). \tag{4.19}$$

Now the counterpart of the first correlation in Eq. (4.17) takes the form

$$\frac{x''_1}{x'_1} = 1 + \frac{1}{p}\frac{\partial p}{\partial x'_1}(1 - x'_1). \tag{4.20}$$

Again, for the counterpart of Eq. (4.18) we have

$$\frac{x''_1}{x'_1} = \left[1 - \frac{1}{p}\frac{\partial p}{\partial x''_1}(1 - x''_1)\right]^{-1}. \tag{4.21}$$

In contrast to the diagram at p = constant, where the calculation of one of the phase equilibrium lines by the other line requires the knowledge of the phase transition heat L, here it is sufficient to know the equation of state of the ideal gas.

In thermodynamics the model of a regular solution is often used along with the model of ideal solutions [226]. The value of the excess entropy s^E, as assumed for this model, is the same as for an ideal solution, i.e., $s^E = -R(x_1 \ln x_1 + x_2 \ln x_2)$, but the excess enthalpy h^E

(or the corresponding heat of mixing, h^M) is different from zero. For one mole of the solution, we have $h^E = qx'_1x'_2$. The model parameter q is related to the activity coefficients γ_1 and γ_2 of the components in the expression for the chemical potential

$$\mu_i = \mu_{0i}(T,p) + RT\ln\gamma_i x_i \tag{4.22}$$

by the following relations

$$RT\ln\gamma_1 = qx_2^2\,, \qquad RT\ln\gamma_2 = qx_1^2\,. \tag{4.23}$$

In such a model, the chemical potentials μ'_1 and μ'_2 of the condensed phase have the form

$$\begin{aligned}\mu'_1 &= \mu'_{01}(T,p) + RT\ln x'_1 + q{x'}_2^2\,,\\ \mu'_2 &= \mu'_{02}(T,p) + RT\ln x'_2 + q{x'}_1^2\,.\end{aligned} \tag{4.24}$$

For the right-hand side of Eq. (4.12),

$$F = x''_1\frac{\partial\mu'_1}{\partial x'_1} + x''_2\frac{\partial\mu'_2}{\partial x'_1}\,, \tag{4.25}$$

we have in this case

$$F = RT\left(1 - \frac{2q}{RT}x'_1x'_2\right)\left(\frac{x''_1}{x'_1} - \frac{x''_2}{x'_2}\right)\,. \tag{4.26}$$

The right-hand side of Eq. (4.15) contains the following factor [228]

$$d = 1 - \frac{2q}{RT}x_1x_2\,. \tag{4.27}$$

For the model of regular solutions, instead of Eq. (4.17), at p = constant, we obtain the following expressions

$$\beta = 1 - b'_p(1 - x'_1)\,, \qquad \alpha = \frac{1 - b'_p(1 - x'_1)}{1 + b'_px'_1}\,, \tag{4.28}$$

where $b'_p = b_p/d$. The transition to the ideal solution is accomplished in the limit $q \to 0$ and $d \to 1$.

4.2 Application of the Generalized Clausius–Clapeyron Equation for the Plot of the Phase Diagrams

A simple liquid–vapor phase diagram of a two-component system at constant pressure, p, and constant temperature, T, is shown in Figs. 4.1a and b. The compositions x'_1 and x''_1, labelled in Fig. 4.1a, are the equilibrium molar fractions of the first component in liquid and vapor at some intermediate temperature between the temperatures of phase equilibrium of pure components at the given pressure.

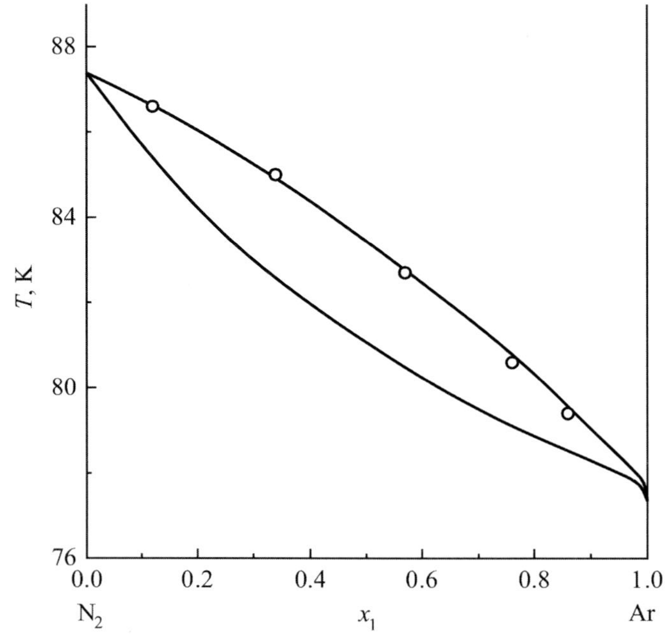

Figure 4.2: Phase diagram of the argon-nitrogen binary system. Full lines refer to experimental data, the dots represent results of computations for the vapor branch in the approximation of a perfect solution.

We consider the following problem. Assume we know one of the lines A or B, then how do we find the other line using Eq. (4.12)? As a first approximation, we can use the model of an ideal solution, where Eqs. (4.17) are obtained for the distribution/separation coefficients. We calculate β or α for the given point on the line A by the known values of the heat of phase transformation, L, and of the derivative $(\partial T/\partial x_1')_p$ and consequently we know the concentration x_{1id}'' on the vapor branch

$$x_{1id}'' = x_1'\beta \qquad \text{or} \qquad x_{1id}'' = \left(1 + \frac{1 - x_1'}{\alpha x_1'}\right)^{-1} . \tag{4.29}$$

This value, x_{1id}'', is marked in Fig. 4.1a. In the case of an ideal solution x_{1id}'' coincides with x_1'' on the line B. A similar result is obtained for the calculation of x_{id}' by using the vapor branch.

The calculation result for the vapor branch of an argon–nitrogen system at atmospheric pressure is compared in Fig 4.2 with the experimental curve (upper solid curve [229]). Here we have a good agreement: the system is close to the ideal one. Figures 4.3 and 4.4 show examples of calculation of the (p–x)-phase diagrams for nitrogen–oxygen and benzene–toluene binary solutions. Both systems are close to ideal ones. The solid lines in the diagrams are plotted using the experimental data given in Ref. [229]. The points correspond to the calculation in two cases: in the first one the change of volume at the phase transition has been determined

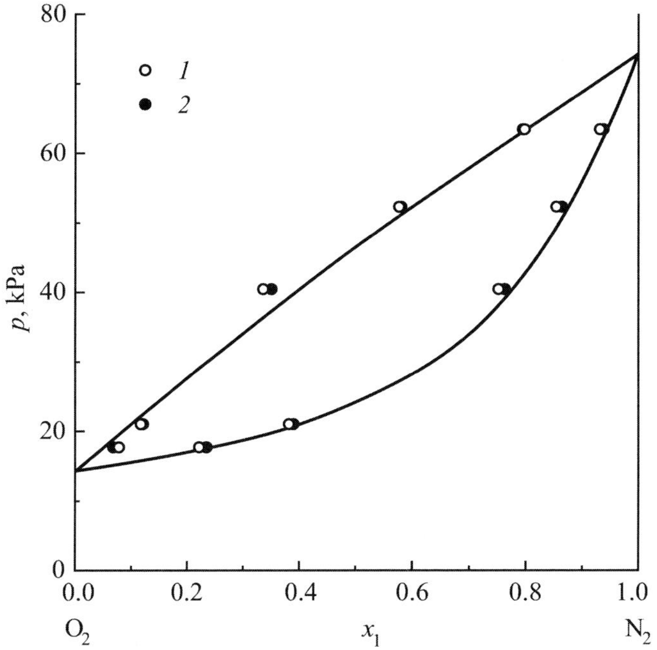

Figure 4.3: $(p–x)$-liquid–vapor phase diagram for the nitrogen–oxygen system at a temperature 75 K. The dots refer to results of computations: (1) results if the change of the volume, Δv, was determined by Eq. (4.11); (2) results of the computations based on Eq. (4.19).

by partial molar volumes of the components using Eq. (4.11); in the second case, Eqs. (4.19) were used for the calculation of the volume jump. In both cases we have a good agreement of the results of the calculations with the experimental data. Figure 4.5 shows results for the nonideal heptane–ethylbenzene system, where the calculation of the vapor branch in the approximation of the ideal solution (open circles) significantly deviates from experimental data [229] shown by the solid line.

Turning back to Eq. (4.12) we see that the natural way of improving the calculation of two-phase equilibrium is to use more precise solution models as compared to the model of an ideal solution. But in this case the results significantly depend on the parameters of the model. Even in the simplest case of the model of regular solutions, described by Eq. (4.23), we have the parameter q, which is related to the excess enthalpy h^E via $h^E = qx_1x_2$, and presumes the existence of the critical temperature T_c of mixing of the components in a condensed phase [226] according to the correlation $h^E/RT = 2(T_c/T)x_1x_2$.

Also an alternative way (see Refs. [223, 230–232]) has been proposed to improve the calculation method employing the model of ideal solutions in Eq. (4.12). In such an approach the ideal solution is taken as a basic reference system for comparison. The deviations of the distribution/separation coefficients in the ideal case from their values in a real system for the same values of T, p, x are related to the values Ω_j of a different nature, which characterize the nonideality of the solution, namely the excess values of enthalpy, volume and entropy. The

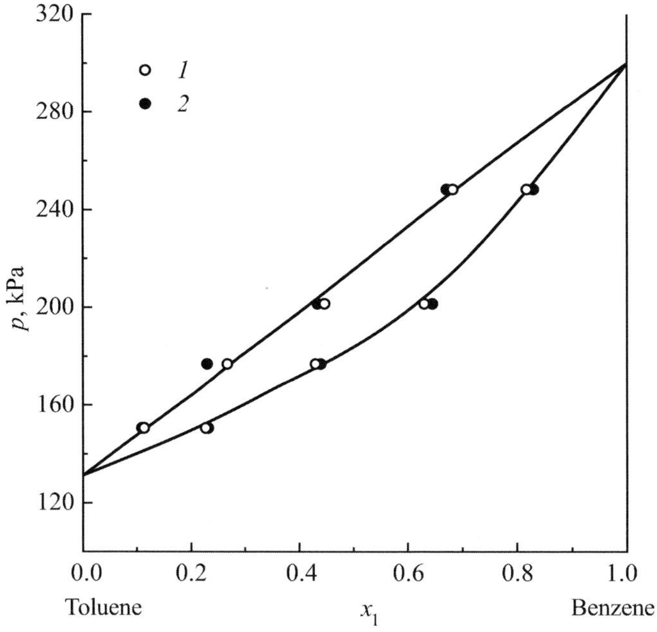

Figure 4.4: (p–x)-liquid–vapor phase diagram for the benzene–toluene system at a temperature of 393.2 K. The dots refer to results of computations (1) if the volume difference, Δv, is determined via Eq. (4.11) or (2) via Eq. (4.19).

existence of the following functional correlation is supposed

$$g = \frac{\alpha_{id}}{\alpha} = f(\Omega_1, \Omega_2, \ldots) . \tag{4.30}$$

The easiest case is when the right-hand side of Eq. (4.30) contains only one parameter, e.g., $\Omega_1 = h^E/RT = \widetilde{h}^E$. For the ideal solution, we have $h^E = 0$.

For binary systems with van der Waals forces of interaction between molecules the relative separation coefficient g and the dimensionless excess solution enthalpy are strongly correlated [230, 231]. In a linear approximation, the following correlation is proposed

$$g = 1.12 - 1.54\widetilde{h}^E , \qquad 0 \leq \widetilde{h}^E \leq 0.6 . \tag{4.31}$$

Note, that in Refs. [230, 231] the inverse value of the separation coefficient was taken in relation to Eq. (4.16), and therefore also for g (Eq. (4.30)).

The phase diagrams used to build the correlation dependence Eq. (4.31) are related to atmospheric pressure. The matching of g and $\widetilde{h}^E$ for each system has been made at one or several points of the line A (Fig. 4.1a). Having the dependence $g = f(\widetilde{h}^E)$ one can correlate the data of the mixing heat of the components and of the separation coefficient for some less-studied system. Having the data for the solution mixing heat and using the correlation

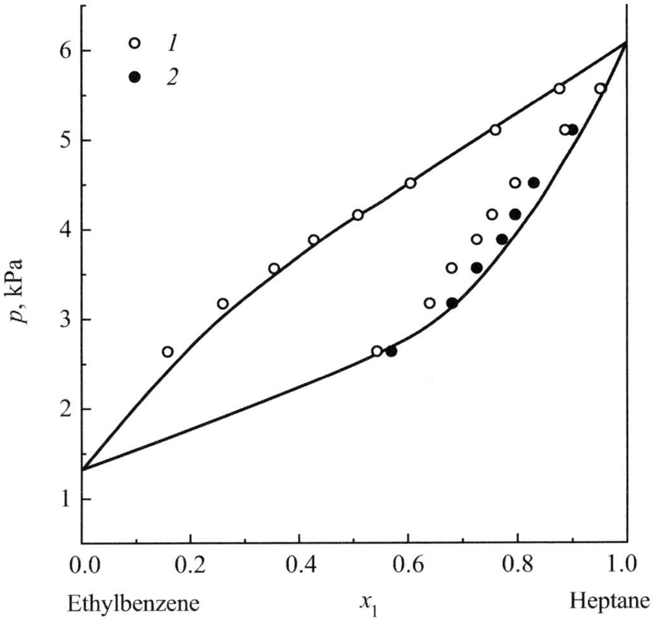

Figure 4.5: Phase diagram of liquid–vapor for the heptane–ethylbenzene binary system at a temperature of 298.2 K. The full curves are drawn according to experimental data. The dots refer to results of computations (1) assuming the system to be a perfect solution and (2) accounting for the nonideality of the solution.

Eq. (4.31) one can introduce the correction for the nonideality of the system and improve the estimate of the vapor branch of the liquid–vapor phase diagram. This procedure can be seen from the plot in Fig. 4.5 made for the equilibrium line of the heptane–ethylbenzene solution.

The correlation Eq. (4.31) should be considered as a first attempt to realize the idea of thermodynamic similarity for the phase equilibrium of solutions. The use of p = constant phase diagrams at atmospheric pressure is not the best choice, as the mixing heats are usually determined at lower pressure, when the system is far from the boiling state. The extrapolation of the values h^E to the boiling temperature of the solution at atmospheric pressure is not reliable. But there is another fundamental reason why the single-parameter similarity correlation, $g = g(h^E)$, is not universal. It is because the measured mixing heat can contain a contribution caused by the chemical interaction of the components, which is assumed to be excluded in the general notion of an ideal solution. Let us demonstrate this for the example of the solution of light and heavy water. When the solution $H_2O - D_2O$ is enriched with the heavy component by the distillation method [233], the system behaves like an ideal solution. While equal amounts of liquid H_2O and D_2O are mixed at 25 °C the temperature decreases by more than 0.3 K. The mixing heat h^E is 7.64 cal/mol, or $\widetilde{h}^E = 0.013$ [234]. This effect should be attributed to the isotope exchange reaction heat $H_2O + D_2O \leftrightarrow 2HDO$ [234]. In the case of other systems with quasichemical interaction of the components it is more difficult to determine the contribution of such interaction.

Generalizing the equation of state $(p + a/v^2)(v - b) = RT$ to the case of two-component systems [225] van der Waals has noted that for "normal" substances the force parameters "a" in this equation satisfies the inequality $a_{12} < \frac{1}{2}(a_{11} + a_{22})$. This statement implies a weakening of the intermolecular interaction during the mixing of the components and is in agreement with $\widetilde{h}^E > 0$. He added further: "The systems where the inequality is not true reveal chemical transformations between components or other deviations from simple substance behavior" [225].

In systems having hydrogen bonds, associates are formed and the correlation between $\widetilde{h}^E$ and g becomes more complicated. For such systems, the correlation Eq. (4.31) does not hold. On the plane $(\widetilde{h}^E, g)$, the data are separated, and the solutions form several groups which follow their particular thermodynamic similarity laws [230, 231]. The formation of the solution can be accompanied by a significant change of the contribution of hydrogen bonds to the energy of the system and to the excess enthalpy of the solution. For many aqueous solutions, the inequality $\widetilde{h}^E < 0$ holds. There are systems which are close to an ideal solution with respect to the distribution of the components between the phases, $g = 1$, at $\widetilde{h}^E \neq 0$. The mixing heat in this case is interpreted [226] as the heat of forming or dissociation of complexes. Our approach reveals such systems very well [230, 231].

Taking into account the above considerations, even the model of an ideal solution should be given a more precise definition. Following Prigogine and Defay [226], the authors base their discussion on the correlation Eq. (4.13): "The system is called ideal, if the chemical potential, μ_i, of each of the components linearly depends on the logarithm of the molar fraction, x_i, where the coefficient in front of $\ln x_i$ is equal to RT. If the system is ideal on the whole domain of concentrations up to $x_i = 1$, then $\mu_i^0(T, p)$ is the chemical potential of the pure i-th component at the given temperature and pressure." These properties are assumed to be fulfilled in our model. Also it is supposed that the relations $h^E = 0$ and $v^E = 0$ hold.

Now we consider further examples of liquid–vapor diagrams of two-component systems. Figure 4.6 shows the phase diagram of the water–dimethylacetamide system at normal atmospheric pressure. The solid lines are plotted based on experimental data [235] for liquid–vapor equilibrium. The squares denote the calculation results of the vapor phase composition in the model of an ideal solution obtained by the known $T(x_1')$-curve for the liquid. The triangles denote the points for the liquid calculated on the basis of the data for the vapor. As follows from Fig. 4.6, for the considered example the correlations Eqs. (4.15), (4.17) and (4.18) give quite good agreement with the experiment for each phase. The heat of mixing of the components in the liquid phase is negative: for 25 °C and $x_1' = 0.5$ it is about -2500 J/mol [235], $\widetilde{h}^E = -1.01$.

The presented approach is applicable also for metal systems having a small mixing heat and forming a continuous series of solid solutions in a crystal phase [236]. The alloys of certain alkali metals are examples of such systems. For a cesium–potassium alloy there are experimental data for the liquid line of the (p, x)-phase diagram at different temperatures [237]. Using the described method, we can calculate the vapor line. The results, obtained for a temperature of 800 K, are given in Fig. 4.7.

Until now we considered liquid–vapor phase equilibria for two-component systems. But Eqs. (4.17) and (4.18) obtained for the case of an ideal solution also hold true for crystal–liquid phase equilibrium, when there is a domain of continuous solid solutions. The phase

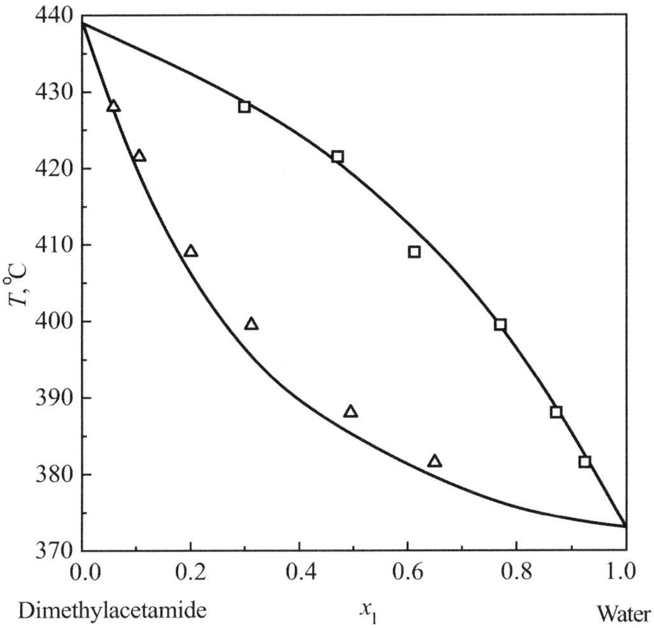

Figure 4.6: (T, x)-phase diagram of the water–dimethylacetamide system at atmospheric pressure. The full curves refer to experimental data, the dots to results of computations employing Eqs. (4.15), (4.17) and (4.18).

transition heat, L, comprises now additively of melting heats of the components L_1 and L_2. If the liquidus line is calculated by the known solidus line (low-temperature phase), then $L = L_1 x_1'' + L_2 x_2''$, where x_1'' and $x_2'' = 1 - x_1''$ are the molar fractions on the liquidus line at a given temperature (see Fig. 4.1a). The concentration x_1'' is not known a priori, so we can use the method of successive approximations to refine the value of L for the calculation of $x_1'' = x_1' \beta$.

Figure 4.8 shows the crystal–melt phase diagram of the cesium–potassium system [238], which has eutectic point. The circles denote the calculated values of x_1'' in the model of ideal solutions for several temperatures [238] at both sides of the temperature minimum. The good agreement of these points with the experimental liquidus line correlates with the low heat of mixing of the components, $\widetilde{h}^E \approx 0.02$ [238], which gives evidence of the low nonideality of the cesium–potassium system in our approach. Additional information about the calculation of the liquidus line of metallic systems can be found in Refs. [231, 238].

The above examples of phase diagram construction using the model of ideal solutions indicate the simplicity and efficiency of the approach. In a number of cases the position of the phase equilibrium line for one of the phases obtained by the line of the other phase using this approximation is close to experiment, while Eqs. (4.17) for ideal solutions are bound to the real correlation for the other phase, e.g., $x_1' = x_1'(T, p)$. For systems with a van der Waals intermolecular interaction there is a correlation similar to Eq. (4.31) between the relative

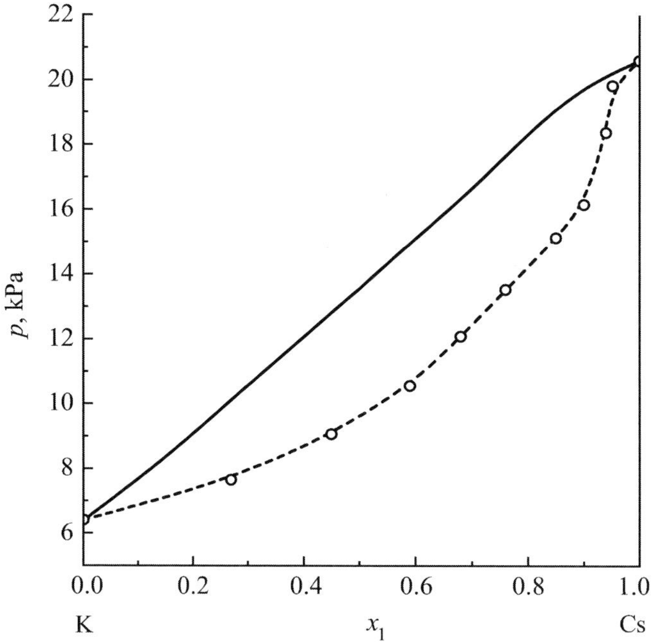

Figure 4.7: (p, x)-liquid–vapor phase diagram for the cesium–potassium system at a temperature of 800 K. The full curve refers to experimental data, the dots connected by the dashed curve refer to computations of the vapor line based on experimental data for the liquid line.

separation coefficient, $g{=}\alpha_{id}/\alpha_{exp}$, and the reduced excess enthalpy, $\widetilde{h}^E > 0$. This feature allows us to introduce a correction to α_{id} and to make the calculation results close to the experimental data for the sought equilibrium line. The proposed application of the model of ideal solutions and the concept of thermodynamic similarity complement the traditional approach for studying solutions.

As an addition to the above discussion we consider the known problem of the temperature shift of the liquid–vapor or liquid–crystal phase equilibrium of a solution for the case when one of the components is a small addition, which is almost not present in the second phase. To be definite, first we consider a liquid–vapor phase equilibrium. Let x_1' be the molar fraction of the main component of the liquid. The content of the other component, which does not evaporate, $x_2' = 1 - x_1' = \varepsilon$ is considered as a small quantity, i.e., $\varepsilon \ll 1$. In the diluted solution we apply the model of the ideal solution to the main component

$$\mu_{01}'(T_1, p) + RT \ln x_1' = \mu_{01}''(T_1, p) + RT \ln x_1'' \,. \tag{4.32}$$

From this correlation we get the distribution of the substance between the phases as (Nernst's rule)

$$\frac{x_1''}{x_1'} = \exp\left(\frac{\mu_{01}' - \mu_{01}''}{RT}\right) \,. \tag{4.33}$$

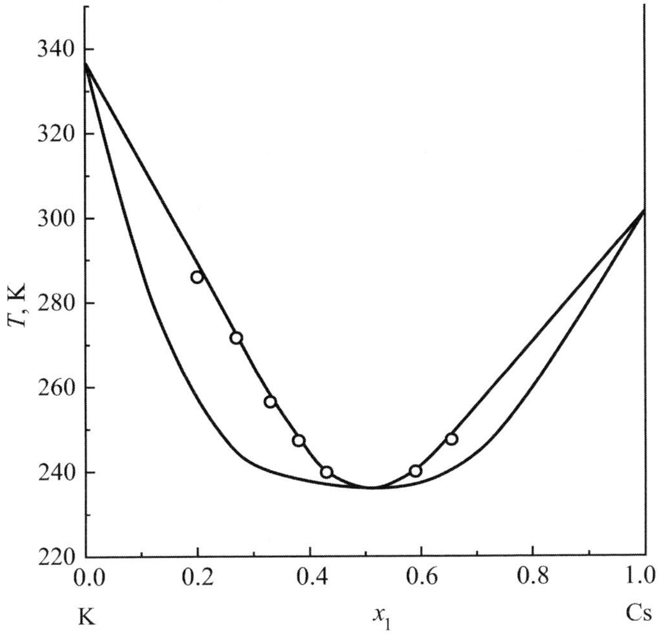

Figure 4.8: Comparison of the experimental phase diagram of the cesium–potassium system (full curves) with results of computations along the liquidus line (circles).

Due to the presence of the additives the chemical potentials $\mu'_{01}(T,p)$ and $\mu''_{01}(T,p)$ differ at the given pressure from their values at a temperature T_0, for which the phases of the pure substance are in equilibrium. We get

$$\mu'_{01}(T_0,p) = \mu''_{01}(T_0,p) \; . \tag{4.34}$$

Writing the differences of chemical potentials in Eq. (4.33) in the form

$$[\mu'_{01}(T,p) - \mu'_{01}(T_0,p)] - [\mu''_{01}(T,p) - \mu''_{01}(T_0,p)] \; , \tag{4.35}$$

expanding the differences in the brackets as series in $\Delta T = T - T_0$ and taking only first-order terms of the expansion, we get for the exponent in Eq. (4.33) the following expression

$$(S''_{01} - S'_{01})\frac{\Delta T}{RT} = \left(\frac{L_1}{RT^2}\right)\Delta T \; , \tag{4.36}$$

where L_1 is the phase transition heat of the main component. When $|(L_1/RT^2)\Delta T| \ll 1$ holds, which is always true for $\varepsilon \ll 1$, we can represent Eq. (4.33) in the following form

$$\frac{x''_1}{x'_1} = 1 + \frac{L_1}{RT^2}\Delta T \qquad \text{or} \qquad \Delta T = \left(\frac{RT_0^2}{L_1}\right)\varepsilon \; . \tag{4.37}$$

The correlation Eq. (4.37), which expresses the increase of the boiling temperature of the solution in the presence of a nonvolatile dopant (where ε is its molar fraction), is called the ebullioscopic formula and the value $E = RT_0^2/L_1$ is called the ebullioscopic constant.

Similarly we can obtain the cryoscopic formula for the decrease of the freezing temperature, $\Delta T = T - T_0$, of the solution in comparison with the pure substance, when the dopant does not go over into the crystal low-temperature phase

$$-\Delta T = \left(\frac{RT_0^2}{L_1}\right)\varepsilon \qquad \text{where} \qquad \varepsilon = x_2'' = 1 - x_1'' \,. \tag{4.38}$$

The change of the sign of $\Delta T = T - T_0$ as compared to Eq. (4.37) is related to the interchange of the roles of high-temperature and low-temperature phases. The value $K = RT_0^2/L_1$ is called the cryoscopic constant. It is interesting that the value of the phase equilibrium temperature shift in Eqs. (4.37) and (4.38) does not depend on the nature of the dopant. This property is related to the assumed universal behavior of the phases in relation to the dopant (the dopant does not go over into one of the phases). Of course, Eqs. (4.37) and (4.38) could be obtained also from Eqs. (4.17) and (4.18) for the phase equilibrium of the ideal solution if we use the approximation $(\partial T/\partial x_1') = \Delta T/(-\varepsilon)$ for the case of the nonvolatile dopant and the approximation $(\partial T/\partial x_1'') = \Delta T/\varepsilon$ for the noncrystallizing dopant.

The Nernst formula Eq. (4.33) shows that to find the distribution of the two components between the phases (even for an ideal system!) one needs to know the differences $\mu_{01}'(T,p) - \mu_{01}'(T_0,p_0)$ and $\mu_{01}''(T,p) - \mu_{01}''(T_0,p_0)$ when the equilibrium is shifted with respect to the equilibrium of the phases of the pure component, when $\mu_{01}'(T_0,p_0) = \mu_{01}''(T_0,p_0)$ holds. In general, these differences are expressed through the integrals taken with respect to T and/or p. For one of the phases, the integrals are related to the metastable states. The simplification which was used to develop the ebullioscopic and cryoscopic formulas is in fact connected with the circumstances that, due to the smallness of the concentration of the dopant, $\varepsilon \ll 1$, we could restrict ourselves to Eq. (4.33), which is related to one (the prevailing) component of the system, and identify the phase transition heat, L, with the phase transition heat of a single-component system. The "local" approach developed in Section 4.1, leads to the use of the known values of entropy and volume of the components in the phases and does not require the introduction of integral contributions to the thermodynamic correlations.

4.3 Thermodynamic Correlations for Phase-Separating Solutions

In the present and the next section we will consider the two-phase liquid–liquid equilibrium in phase-separating two-component systems having an upper or a lower critical dissolution temperature. Equation (4.12) holds true also for this case, but here we cannot use the model of ideal solutions. The possible simplification is the application of Eq. (4.12) for the isopleths $x_1'(T,p) = \text{constant}$, $dx_1' = 0$ near the critical dissolution point, $x_1' \cong x_{1c}(T,p)$. Assuming a weak dependence of x_{1c} on T and p we obtain Eq. (4.12), where the right-hand side is equal to zero. Now the formally generalized Clausius–Clapeyron equation has the ordinary form, i.e., $(dp/dT) = (L/T\Delta v)$.

The equilibrium of the two-phase liquid system is defined by equality of the chemical potentials in both phases. For the first component we have

$$\mu_1'(T,p,x_1') = \mu_1''(T,p,x_1'') \,. \tag{4.39}$$

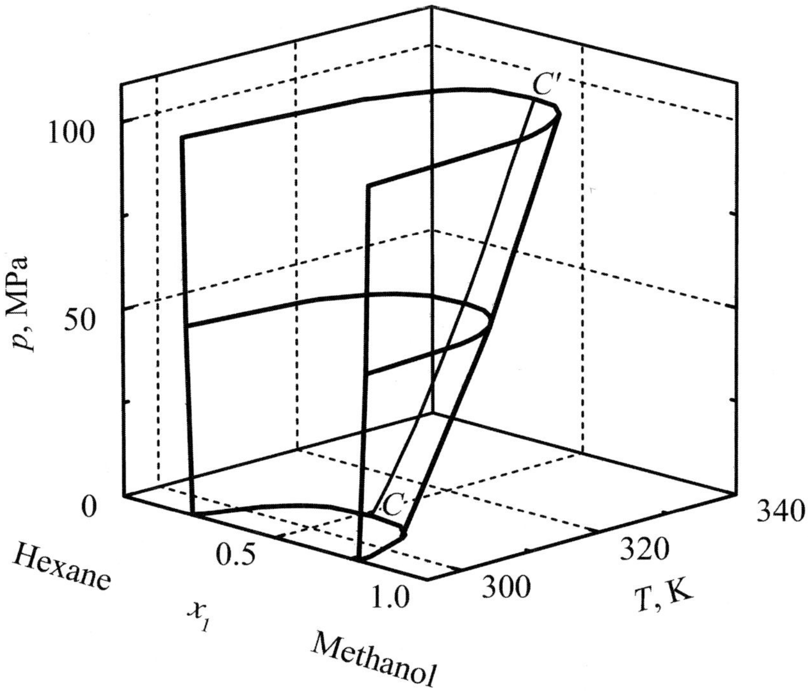

Figure 4.9: (T, p, x)-diagram of the liquid–liquid phase equilibrium for the methanol–n-hexane system. (CC$'$) is the line of critical points of the solution.

Here the values denoted with one and two primes are related respectively to the first and the second phase. The stability of the phases with respect to small (homophase) perturbations is ensured by the condition

$$\left(\frac{\partial \mu_1}{\partial x_1}\right)_{T,p} > 0 \,. \tag{4.40}$$

When the derivative in Eq. (4.40) tends to zero, it means that the system reaches the spinodal (the boundary of stability of the homogeneous solution with respect to small concentration fluctuations)

$$\left(\frac{\partial \mu_1}{\partial x_1}\right)_{T,p} = 0 \,. \tag{4.41}$$

This equation and the conditions

$$\left(\frac{\partial^2 \mu_1}{\partial x_1^2}\right)_{T,p} = 0 \,, \tag{4.42}$$

$$\left(\frac{\partial^3 \mu_1}{\partial x_1^3}\right)_{T,p} > 0 \tag{4.43}$$

define the critical dissolution point. The solution spinodal, which is defined by the condition Eq. (4.41) at $\left(\partial^2\mu_1/\partial x_1^2\right) \neq 0$, is sometimes called the diffusion spinodal in contrast to the "mechanical" spinodal, where $(\partial p/\partial v)_T = 0$, $\left(\partial^2 p/\partial v^2\right)_T \neq 0$ holds. In the general case, there are two kinds of boundaries of stability of the homogeneous state of the solutions [226].

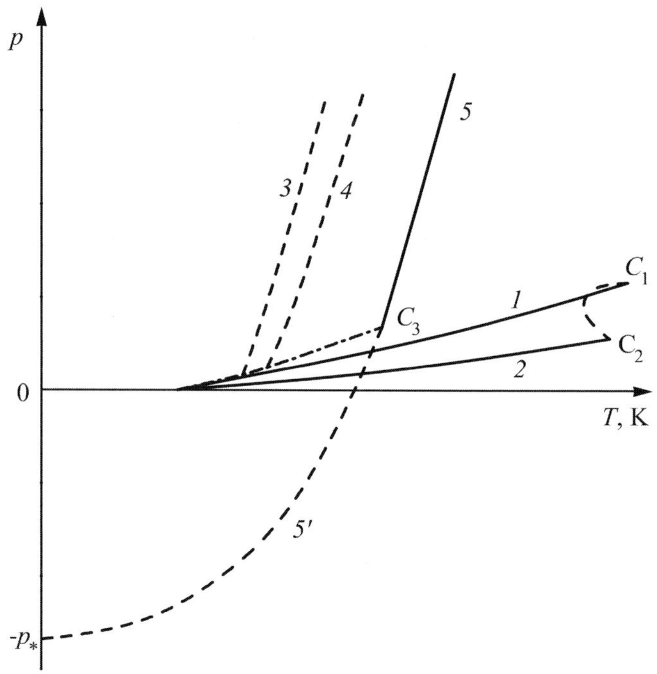

Figure 4.10: Phase diagram of a binary solution with upper critical dissolution point.

It has been noticed that the critical isopleths $x_{1c}(T,p)$ of systems with an upper critical dissolution temperature on the (T,p)-plane for positive values of excess volume and enthalpy look similar to the melting lines and can be approximated by the Simon equation Eq. (3.3). A typical example of this class of systems is the methanol–n-hexane system, the (T,p,x)-equilibrium phase diagram of which is shown in Fig. 4.9 [239]. The region of homogeneous states of the solution is located on the diagram at temperatures above the surface of phase equilibrium. The (CC′) line describes the change of the critical temperature with the change of pressure.

The coexistence curve (binodal of the solution) in the vicinity of the critical temperature T_c at fixed pressure is described by the scaling correlation

$$\frac{x_1'' - x_1'}{2x_{1c}} = B\left|\frac{T_c - T}{T_c}\right|^{\beta} \tag{4.44}$$

with the universal index $\beta \approx 0.325$. Here B is an individual constant (binodal amplitude) depending on the properties of the components.

The problem of phase equilibrium in solutions has an important facet: the possibility to extend the phase equilibrium line into the region where both coexisting phases are metastable. Figure 4.10 shows the (p, T)-phase diagram of the system with upper critical dissolution temperature. Here lines 1 and 2 are the pressure curves of the saturated vapor of the pure components; 3 and 4 are the melting lines of the substances. The dashed line between the critical points C_1 and C_2 is the line of liquid–vapor critical points for the solutions. Line 5 of liquid–liquid critical points meets at point C_3 with the line of three-phase liquid–liquid–vapor equilibrium (dashed-dotted line). The segment $5'$ is the metastable extension of line 5 beyond point C_3. Both coexisting liquid phases below C_3 are metastable with respect to the vapor phase, but if the latter is absent, their equilibrium is possible, if the chemical potentials of the phases, Eq. (4.39), are equal and each phase retains stability with respect to small (homophase) perturbations. The indirect confirmation of the reality of such extensions are the numerous experiments [1] with superheated (stretched) liquids.

Line 5 in Fig. 4.10 is extended beyond the point C_3 using the Simon approximation Eq. (3.3). The asymptotic pressure $-p_*$ at $T = 0$ has the same physical meaning as in the case of the metastable extension of melting lines of single-component substances. The value p_* can be the scaling parameter of the pressure for the condensed phase. As an order of magnitude, p_* corresponds to the internal pressure in the liquid $(\partial u/\partial v)_{T\to 0}$, or the theoretical tensile strength of the liquid. For instance, for the methanol–n-hexane system $p_* = 90.3$ MPa. According to van der Waals, the maximum stretching pressure the liquid can sustain (it is located on the spinodal) is $27p_c$ at $T = 0$. For hexane $27p_c = 81.8$ MPa, which is close to the above mentioned value of p_*. For methanol $27p_c = 218$ MPa.

4.4 Experimental Studies of Phase-Separating Solutions

The first study of the influence of pressure on the critical dissolution parameters was made by Timmermans in the 1920s [240]. Several tens of solutions with upper and lower critical dissolution temperatures were studied for pressures up to 100 MPa. This investigation showed the diversity in the behavior of the critical curves. The experiment was carried out using the visual observation of the solutions in the region of the phase transformation. In the experiments the turbidity of the critical solutions was observed which preceded the decomposition. Due to its simplicity and convenience the method of measuring the (p, T)-curves of liquid–liquid equilibrium, which is based on the determination of temperatures of solution turbidity at the point of the phase transformation (Alexeev method [241]), is very common for studying phase-separating liquid solutions, which are transparent for visible radiation.

References [242–248] are devoted to the investigation of the influence of pressure on the decomposition temperature of the solutions for a large group of binary systems with both upper and lower critical dissolution temperatures. The experiments employed a high-pressure chamber, as shown in Fig. 4.11 [242, 244]. The chamber had a window of molten quartz and allowed one to make visual observations of the sample in the pressure range from atmospheric pressure up to 230 MPa. The pressure-transmitting medium was separated from the studied liquid with the sylphon (4). The pressure was created with a piston pressure gauge (MP-2500) with an accuracy grade of 0.05. The temperature was measured with a copper–constantan thermocouple, which was located directly inside a high-pressure chamber. The precision of

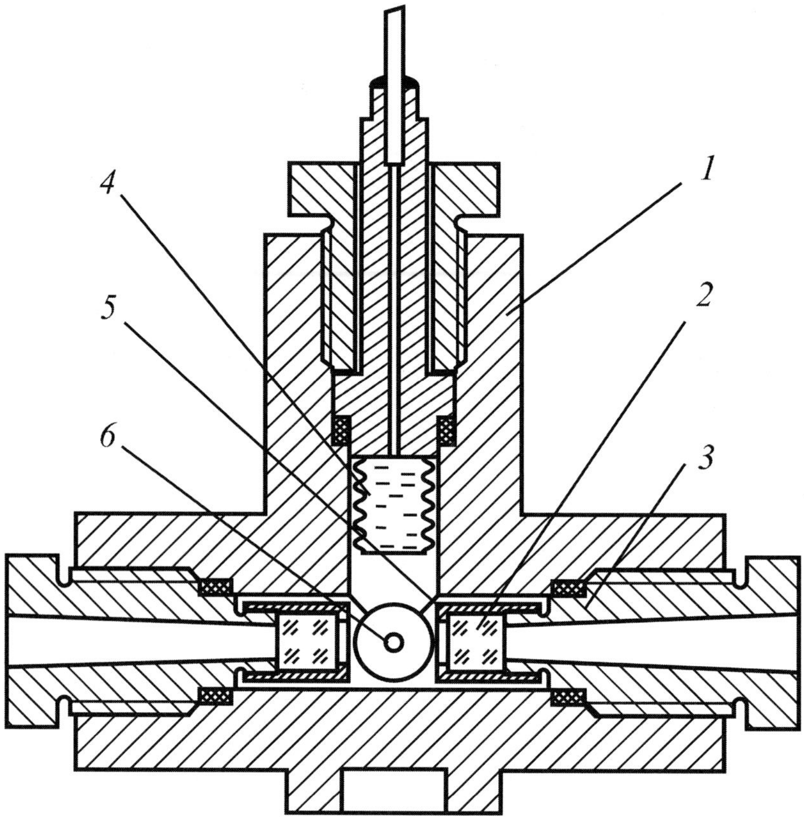

Figure 4.11: High-pressure chamber: (1) steel bomb, (2) window made of molten quartz, (3) optical obturator, (4) sylphon, (5) cap, (6) thermocouple.

temperature measurement in the experiment was about 0.05 K. The chamber was put into the liquid thermostat, which allowed the realization of a temperature in the range from 200 K to 420 K with a precision of ± 0.01 K. The weight methods were used to prepare the solution. The composition of the sample was determined with a precision not lower than 0.02%. The camera was filles with the studied solution at a temperature corresponding to the homogeneous state of the sample. The experiments for the solutions with upper critical dissolution temperature were carried out at constant temperature, while the pressure was gradually increased. The systems with lower critical dissolution temperature were studied at fixed pressure, while the temperature in the thermostat was increased up to the decomposition of the solution at a rate of ~ 1 K/min. The phase transformation point was established by the turbidity of the sample, which preceded the decomposition.

The experimental data for the dependence of decomposition temperature on pressure for compositions close to critical ones of two methanol–alkane solutions and three acetone–alkane solutions are presented in Fig. 4.12a. The data for five alkane-fluorocarbon systems and for carbon tetrachloride-perfluoromethylcyclohexane are shown in Fig. 4.12b. The critical

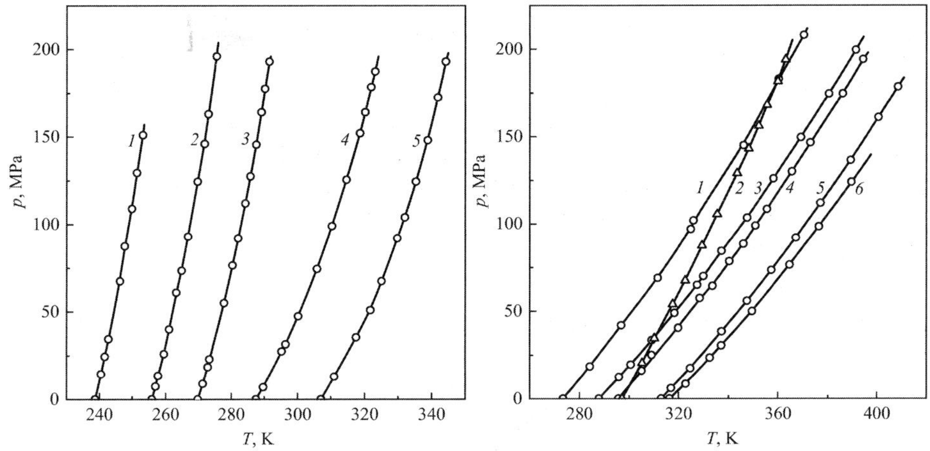

Figure 4.12: Dependence of the critical dissolution temperature on pressure for binary liquid solutions with upper critical dissolution point: (a) (1) acetone–n-hexane, (2) acetone–n-octane, (3) acetone–n-decane, (4) methanol–n-pentane, (5) methanol–n-hexane; (b) (1) n-pentane–perfluorohexane, (2) tetrachloromethane–perfluoromethylcyclohexane, (3) n-pentane–perfluorooctane, (4) n-hexane–perfluorohexane, (5) n-octane–perfluoromethylcyclohexane, (6) n-heptane–perfluorohexane.

isopleths were approximated using the Simon equation. In this case the temperature T_0 in Eq. (3.3) corresponds to the critical dissolution temperature at zero (atmospheric) pressure. The results of the computation of the parameters p_* and c in the equation, the mean quadratic deviations $\sigma(p)$ of the experimental points from the approximating dependence, the critical concentrations of the second component, x_{2c}, and critical temperatures, T_c, at atmospheric pressure are given in Table 4.1.

Table 4.1: Parameters of the Simon equation for the critical decomposition curves for a group of solutions with upper critical dissolution temperature.

Solution	x_{2c}	T_c, K	p_*, MPa	c	$\sigma(p)$, MPa
Methanol–n-pentane	0.517	287.5	98.7	9.33	0.5
Methanol–n-hexane	0.450	307.0	90.3	10.1	0.3
Acetone–n-hexane	0.518	238.7	170.4	10.9	0.6
Acetone–n-octane	0.432	255.9	173.5	10.2	1.2
Acetone–n-decane	0.289	269.9	156.6	10.5	0.8
n-Pentane–perfluorohexane	0.330	273.3	166.5	2.67	1.2
n-Pentane–perfluorooctane	0.266	287.9	171.3	2.52	0.3
n-Hexane–perfluorohexane	0.351	295.8	204.5	2.32	1.3
n-Heptane–perfluorohexane	0.384	316.4	150.1	2.90	0.4
n-Octane–perfluoromethylcyclohexane	0.440	313.0	181.6	2.57	0.3
Carbon tetrachloride–perfluoromethylcyclohexane	0.286	297.2	313.3	2.39	0.3

Figure 4.13 shows the phase equilibrium surface of the water–2,6-dimethylpyridine system plotted by experimental data in the region of the lower critical dissolution temperature line. The critical point lines for systems of this kind have a positive slope at normal pressure passing through the temperature maximum into the region of high pressures. The values of excess volumes and enthalpies close to the lower critical dissolution temperature at atmospheric pressures of similar solutions have negative values. The saddle point on the (T, p, x)-surface corresponds to the double critical point of the solution. Here the lines of lower and upper critical pressures meet.

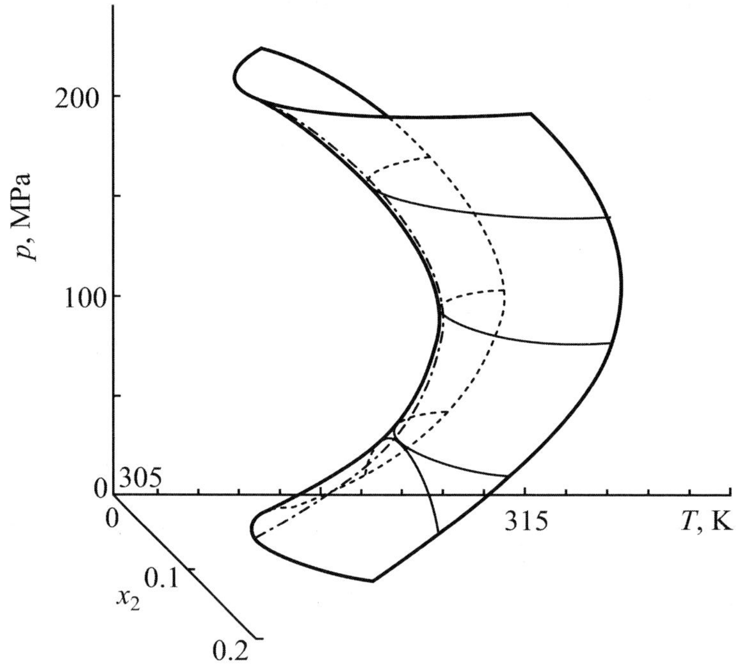

Figure 4.13: The surface of water–2,6-dimethylpyridine solution near the line of the lower critical dissolution points, bounded by the isopleths $x_2 = 0.028$ and $x_2 = 0.125$ (x_2 is the molar fraction of 2,6-dimethylpyridine). The dashed-dotted curve is the critical curve. Some sections are also shown corresponding to $p =$ constant and $T =$ constant.

A significant contribution to the development of experimental investigations of phase-separating solutions at high pressures was made by Schneider. He managed to expand the pressure range up to 600 MPa in experiments with traditional high-pressure chambers in conditions of hydrostatic compression (see, for instance, [249–251]) and up to 3 GPa in experiments with diamond anvils [252]. The investigations of Schneider involve a broad range of substances of different nature: solutions of cryogenic liquids and molecular compounds, water solutions of acids and organic substances. He was the first who classified the experimental data and defined different kinds of behavior of critical dissolution point lines [253].

The record values of pressures in the investigations of solutions were obtained in diamond anvils. Reference [254] contains the study of the influence of the pressure on the decomposition of the helium–hydrogen system for pressures up to 7.5 GPa, and Ref. [255] contains the measurement results for the critical curve of the helium–nitrogen system for pressures up to 10 GPa.

In the last few decades with the development of the experimental techniques the interest of researchers has turned to metallic liquid solutions. But the experiments here are concentrated in most cases on investigations at atmospheric pressure. Metallic systems require new investigation techniques. Among them one can note the method of measuring the electrical resistance [256, 257], which is based on the determination of the phase transformation point by the evolution of the conductivity gradient measuring the height of the liquid column at the moment of the decomposition of the metallic solution. Also there are acoustic methods [258, 259], where the decomposition point is determined by the jump of the velocity of ultrasound on the layer boundary, and the method of gamma-irradiation of the liquid metallic samples [260, 261], where the phase transition is determined by the kink of the dependence of the radiation passing through the sample on temperature. The latter is similar to the method of light scattering, which is widely used for studying solutions of molecular liquids, polymers and water mixtures by the sharp change of the light scattering (absorption) intensity at the transformation point [262, 263].

In Ref. [264] the resistance method is used for studying the dependence of the critical dissolution temperature of five metallic liquid solutions at pressures up to 2.8 GPa. In Ref. [265] the decomposition of the Bi–$BiBr_3$ system for pressures up to 1.7 GPa is studied with the resistance method in a wide range of concentrations.

4.5 Thermodynamic Similarity of Phase-Separating Binary Solutions with Upper Critical Dissolution Temperature

The success in establishing the one-parameter thermodynamic similarity for the melting lines of single-component substances using the extension of the equilibrium curves into the region of metastable states of the coexisting phases allows us to expect that such an approach can be implemented also for studying phase-separating solutions. The one-parameter thermodynamic similarity with respect to the dependence of decomposition temperature on pressure implies the existence of a generic correlation like [244, 245]

$$F\left(\widetilde{p}_{LL}, \widetilde{T}_{LL}, \left(\frac{x_2}{x_{2c}}\right) \approx \text{constant},\ c\right) = 0 \tag{4.45}$$

for the equilibrium lines of a group of substances, where $\widetilde{p}_{LL}$ and $\widetilde{T}_{LL}$ are the reduced pressure and the temperature of liquid–liquid equilibrium and c is an individual parameter of the substance. Let us define the similarity parameter c (analogous to the similarity parameter for the melting lines) through the reduced slope of the equilibrium line at the decomposition temperature T_0, which corresponds to zero pressure

$$c = \frac{T_0}{p_*}\left(\frac{dp}{dT_{LL}}\right)_{T_0} . \tag{4.46}$$

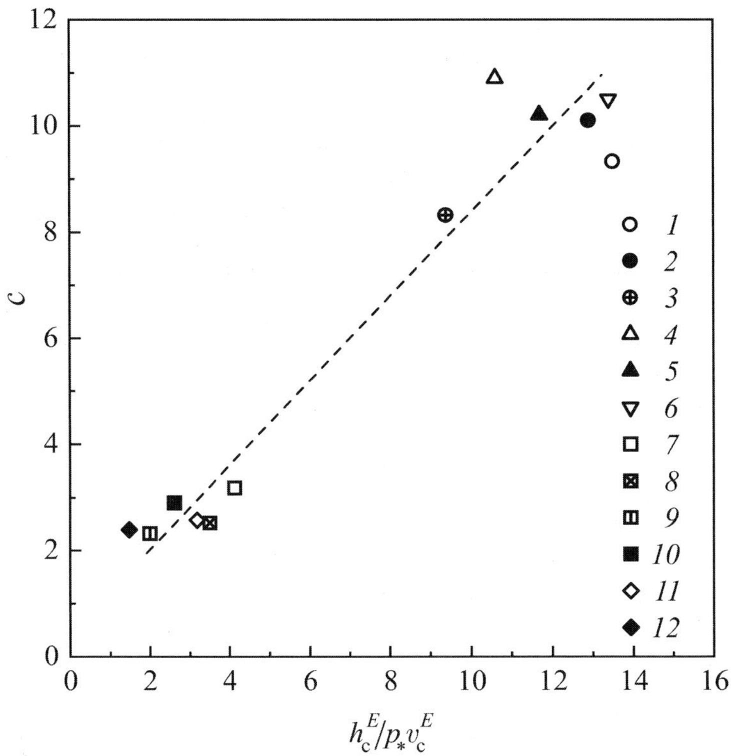

Figure 4.14: Correlation of the similarity parameter, c, with the complex parameter, $(h_c^E/p_* v_c^E)$, for a group of substances with upper critical dissolution point: (1) methanol–n-pentane, (2) methanol–n-hexane, (3) methanol–cyclohexane, (4) acetone–n-hexane, (5) acetone–n-octane, (6) acetone–n-decane, (7) methane–perfluoromethane, (8) n-pentane–perfluorooctane, (9) n-hexane–perfluorohexane, (10) n-heptane–perfluorohexane, (11) n-octane–perfluoromethylcyclohexane, (12) tetrachloromethane–perfluoromethylcyclohexane.

The choice of the Simon equation

$$\frac{p}{p_*} + 1 = \left(\frac{T}{T_0}\right)^c \tag{4.47}$$

as a correlation in Eq. (4.45) assumes that the reduced values of temperature and pressure are used in the form

$$\widetilde{p}_{LL} = \frac{p + p_*}{p_*} , \qquad \widetilde{T}_{LL} = \frac{T}{T_0} . \tag{4.48}$$

It is more difficult to introduce dimensionless variables for enthalpy and volume. If we consider the Clausius–Clapeyron equation for the isopleth $(x_2 \approx x_{2c})$

$$\frac{dp}{dT_c} = \frac{\Delta h_{LL}}{T \Delta v_{LL}} , \tag{4.49}$$

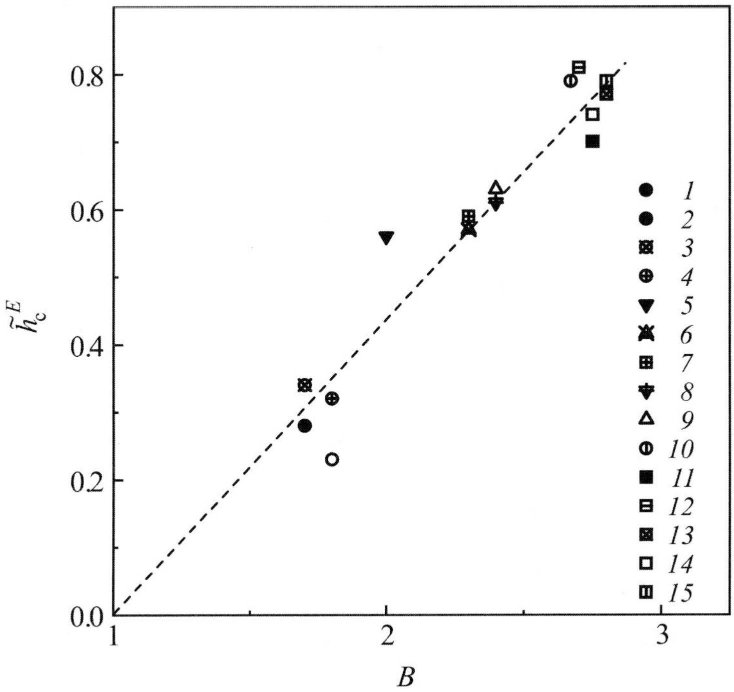

Figure 4.15: Correlation between the amplitude of the binodal curve, B, and the reduced excess enthalpy of the solution of critical composition, $\widetilde{h}_c^E$, for a group of solutions with upper critical dissolution point: (1) methanol–n-pentane, (2) methanol–n-hexane, (3) methanol–n-heptane, (4) methanol–cyclohexane, (5) n-hexane–nitrobenzene, (6) acetone–n-propane, (7) acetone–n-butane, (8) acetone–n-pentane, (9) acetone–n-hexane, (10) n-hexane–perfluorohexane, (11) n-heptane–perfluorohexane, (12) n-octane–perfluorohexane, (13) n-pentane–perfluorooctane, (14) n-hexane–perfluorooctane, (15) n-heptane–perfluorooctane

then with approach to the critical temperature we have $\Delta h_{LL} \to 0$, $\Delta v_{LL} \to 0$ for each pressure, whereas the derivative (dp/dT) retains a certain finite value. The values Δh_{LL} and Δv_{LL} and their variation are difficult to determine experimentally. The authors of Ref. [266] propose introducing into Eq. (4.49) instead of Δh_{LL} and Δv_{LL} the excess enthalpy h_c^E, the volume v_c^E of the solution in the critical point and to use the approximate correlation

$$\frac{dp}{dT_c} \approx \frac{h_c^E}{T_c v_c^E} \,. \tag{4.50}$$

Approximating the transcritical isopleths $p = p(T, x_2 \approx x_{2c})$ with the Simon equation and taking into account Eq. (4.49) we obtain the known correlation of the similarity parameter c with the jumps of enthalpy and volume at phase transition and zero pressure

$$c = \frac{\Delta h_{LL}}{p_* \Delta v_{LL}} \,. \tag{4.51}$$

If the approximation Eq. (4.50) is acceptable, then on the right-hand side of the previous expression we can make the corresponding substitution, i.e.,

$$c \cong \frac{h_c^E}{p_* v_c^E} \,. \tag{4.52}$$

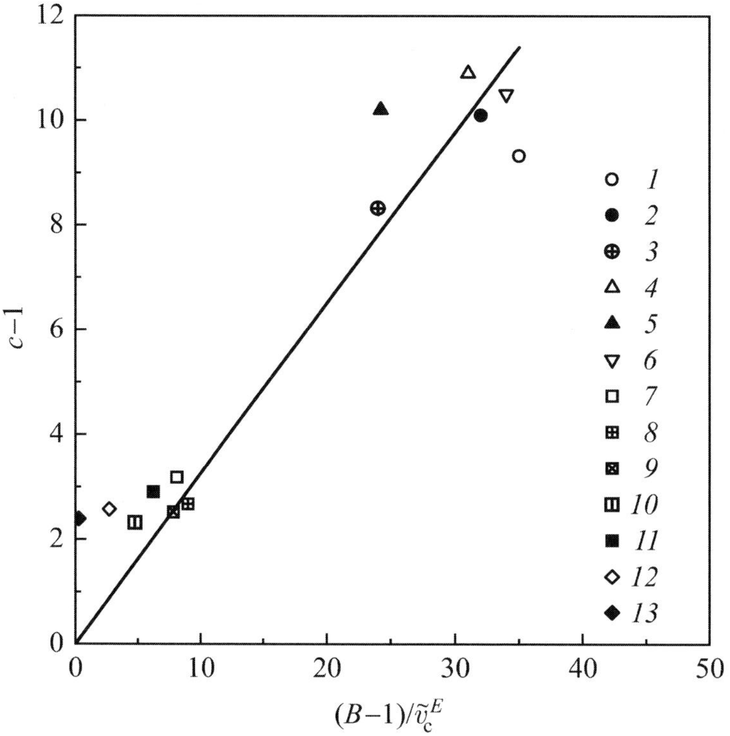

Figure 4.16: Correlation between the similarity parameter, c, and the complex $\left[(B-1)/\tilde{v}_c^E\right]$ for a group of solutions with upper critical dissolution point: (1) methanol–n-pentane, (2) methanol–n-hexane, (3) methanol–cyclohexane, (4) acetone–n-hexane, (5) acetone–n-octane, (6) acetone–n-decane, (7) methane–perfluoromethane, (8) n-pentane–perfluorohexane, (9) n-pentane–perfluorooctane, (10) n-hexane–perfluorohexane, (11) n-heptane–perfluorohexane, (12) n-octane–perfluoromethylcyclohexane, (13) tetrachloromethane–perfluoromethylcyclohexane.

Figure 4.14 shows the correlation between c and the complex $(h_c^E/p_* v_c^E)$. This result can be considered as an indirect confirmation of the correlation Eq. (4.50), while the exponent in the Simon equation (the similarity parameter c) is determined directly from experimental (p, T)-data for the isopleth.

Let us define the dimensionless scaling values of volume and enthalpy as follows:

$$\tilde{v}^E = \frac{v^E p'}{RT} \,, \qquad \tilde{h}^E = \frac{h^E}{RT} \,, \tag{4.53}$$

Table 4.2: Main parameter values characterizing the thermodynamic similarity of critical curves for solutions with upper critical dissolution temperature. x_{2c} is the critical concentration of the second component: (1) methanol–n-pentane; (2) methanol–n-hexane; (3) methanol–cyclohexane; (4) acetone–n-hexane; (5) acetone–n-octane; (6) acetone–n-decane; (7) methane–perfluoromethane; (8) n-pentane–perfluorohexane; (9) n-pentane–perfluoroctane; (10) n-hexane–perfluorohexane; (11) n-heptane–perfluorohexane; (12) n-octane–perfluoromethylcyclohexane; (13) carbon tetrachloride–perfluoromethylcyclohexane.

Solution	x_{2c}	T_0, K	p_*, MPa	c	$\widetilde{v}_c^E$	$\widetilde{h}_c^E$	B
(1)	0.517	287.5	98.7	9.33	0.014	0.19	1.5 [271]
(2)	0.450	307.0	90.3	10.1	0.019	0.25	1.6 [271]
(3)	0.501	318.5	109.2	8.32	0.029	0.30	1.7 [249]
(4)	0.518	238.7	170.4	10.9	0.058	0.61	2.8 [272]
(5)	0.432	255.9	173.5	10.2	0.060	0.70	2.5 [272]
(6)	0.289	269.9	156.6	10.5	0.050	0.66	2.7 [273]
(7)	0.43	94.5	175.0	3.18	0.16	0.66	1.3 [268]
(8)	0.330	273.3	166.5	2.67	0.20	0.77	2.5 [267]
(9)	0.266	287.9	171.3	2.52	0.23	0.77	2.6 [267]
(10)	0.351	295.8	204.5	2.32	0.38	0.75	2.5 [267]
(11)	0.384	316.4	150.1	2.90	0.29	0.76	2.6 [267]
(12)	0.440	313.0	181.6	2.57	0.22	0.70	1.6 [245]
(13)	0.286	297.2	313.3	2.39	0.32	0.49	1.1 [274]

where $p' = p + p_*$. Reference [267] establishes the correlation between the binodal amplitude B in the scaling correlation Eq. (4.44) and the excess enthalpy of the solution of critical composition $\widetilde{h}_c^E$ at atmospheric pressure

$$\widetilde{h}_c^E \sim B - 1 \tag{4.54}$$

(see Fig. 4.15). Now we can establish the correlation between the parameter c and the amplitude of the isobar B by excluding the quantity h_c^E from Eq. (4.52). For a number of systems with hydrocarbons and fluorocarbons having an upper critical dissolution temperature this correlation is shown in Fig. 4.16.

In addition to the solutions studied in the present work in the construction of correlations in Figs. 4.14-4.16 we used also the data for critical point lines of methane–perfluoromethane [268] and methanol–cyclohexane [249] systems. The values of the binodal amplitude B were calculated by experimental data taken from the literature. The references to the corresponding works are given in Table 4.2, which contains along with the parameters of the Simon equation and the values of B the dimensionless excess volumes and enthalpies in the critical points for the studied solutions. The values of $\widetilde{v}_c^E$ of the methane–perfluoromethane solution were calculated using the experimental data for v^E from Ref. [269]; for the other solutions the values of v_c^E were taken from Ref. [245]. The values of $\widetilde{h}_c^E$ were calculated using the data for the mixing heat from Ref. [267], except for the methane–perfluoromethane solution, the data for which were taken from Ref. [270].

As follows from Figs. 4.14 and 4.16, for critical isopleths of solutions with upper critical dissolution temperature the parameter c is strongly connected with the volume effect v^E of solution formation. Like in the case of the melting lines of single-component systems the method of reducing in Eq. (4.53) the parameter v^E using the characteristic internal pressure p_* of the condensed system is of considerable importance for establishing the similarity of decomposing solutions. With the established correlations one can better conceive the relations between various physical parameters in phase-separating liquid solutions. Moreover, they can be used for preliminary estimates of unknown parameters in the established correlations by other known parameters.

4.6 Thermodynamic Similarity of Phase-Separating Binary Solutions with Lower Critical Dissolution Temperature

Phase diagrams with lower critical dissolution temperature are typical for water solutions of organic and heteroorganic compounds. For such systems the transition to the homogeneous state with decrease of temperature is caused by the formation of hydrogen bonds between the molecules of the different components. The critical temperature lines of solutions with lower critical dissolution temperature often have a double critical point, which corresponds to the extremum of the critical curve. The peculiarity of the double critical point is revealed in the behavior of various physical parameters near the double critical point. The kinetic and thermodynamic properties have here evident singularities [246–248, 275–277].

For the solutions with a temperature maximum on the critical curves (see Fig. 4.13) the transcritical isopleths $p = p(T, x_2 \approx x_{2c})$ near the double critical point can be described by the exponential dependence [278]

$$\frac{p_c'' - p_c'}{p_D} = 2C \left| \frac{T_D - T_c}{T_D} \right|^{\lambda} \tag{4.55}$$

with the index $\lambda = 0.5$. Here p_c'' and p_c' are respectively the upper and lower critical pressures at the temperature T_c; p_D and T_D are the pressure and temperature in the double critical point, C is the amplitude of the critical curve.

The double critical point can serve as the reference point for the description of the thermodynamic similarity of such solutions. Using the values p_D and T_D as scales of pressure and temperature one can represent the main thermodynamic properties of the solutions in a dimensionless form. In reduced variables

$$\widetilde{T}_c = \frac{T_D - T_c}{T_D}\,, \qquad \widetilde{p}_c = \frac{p_c - p_D}{p_D}\,, \tag{4.56}$$

the correlation Eq. (4.55) for the critical curve will have the following form

$$\frac{\widetilde{p}_c'' - \widetilde{p}_c'}{2} = C\widetilde{T}_c^{\lambda}\,, \tag{4.57}$$

where $\widetilde{p}_c''$ and $\widetilde{p}_c'$ are respectively the upper and the lower reduced critical pressures at the temperature T_c. The constant C has the meaning of the thermodynamic similarity parameter for the critical point lines of different solutions [279]. To a good approximation the rule of the rectilinear diameter holds true for the critical curves

$$\frac{p_c'' + p_c'}{2} = (Dp_D)\widetilde{T}_c + p_D \ , \tag{4.58}$$

where D is a constant. Equations (4.57) and (4.58) give the following correlation for the critical curve

$$\widetilde{p}_c = \pm C\widetilde{T}_c^{\lambda} + D\widetilde{T}_c \ . \tag{4.59}$$

The "plus" sign corresponds to pressures $p_c > p_D$, the "minus" sign corresponds to the pressures $p_c < p_D$.

The dimensionless values of the excess molar volume and enthalpy can be defined in the form

$$\widetilde{v}^E = \frac{v^E p_D}{RT_D} \ , \qquad \widetilde{h}^E = \frac{h^E}{RT_D} \ , \tag{4.60}$$

correspondingly. The reduced slope of the critical point line at zero pressure

$$\left(\frac{d\widetilde{p}_c}{d\widetilde{T}_c}\right)_{p=0} = \frac{T_D}{p_D}\left(\frac{dp_c}{dT}\right)_{p=0} \tag{4.61}$$

can be taken as the similarity parameter for the critical curves. According to Eq. (4.59) the derivative $(T_D/p_D)(dp_c/dT)_{p=0}$ is correlated with the constant C by the following expression

$$\frac{T_D}{p_D}\left(\frac{dp_c}{dT}\right)_{p=0} = \frac{T_D}{T_D - T_0} - \frac{C}{2}\left(\frac{T_D}{T_D - T_0}\right)^{0.5} \tag{4.62}$$

or in the reduced form

$$\left(\frac{d\widetilde{p}_c}{d\widetilde{T}_c}\right)_{p=0} = \widetilde{T}_0^{-1} - \frac{C}{2}\widetilde{T}_0^{-0.5} \ . \tag{4.63}$$

Here T_0 is the critical temperature at $p = 0$ and $\widetilde{T}_0$ is the corresponding reduced temperature. For practical calculations instead of T_0 one can use the critical temperature T_c at atmospheric pressure, for

$$\frac{|T_c(p = 0) - T_c(p = 1\ \text{atm})|}{T_c(p = 0)} \ll 1 \ . \tag{4.64}$$

The consideration of the thermodynamic similarity in the selected scales of temperature, T_D, and pressure, p_D, in the double critical point leads to the discovery of a number of correlations between dimensionless thermodynamic complexes [279]. These correlations describe a certain class of decomposing systems in reduced variables. In Ref. [279] water solutions of organic compounds were chosen for studying the thermodynamic similarity of phase-separating systems. For such systems the phase equilibrium surface in the vicinity of the lower critical

Table 4.3: Main properties characterizing the decomposition of solutions with lower critical dissolution temperature at atmospheric pressure. x_{2c} is the molar critical concentration of the second component.

Solution	x_{2c}	T_c, K	B	h_c^E, J/mol	v_c^E, cm^3/mol
H_2O–2,4-dimethylpyridine	0.056	295.7	3.0 [282]	−500	−0.34
H_2O–2,6-dimethylpyridine	0.064	307.3	4.3 [246]	−670	−0.44
H_2O–triethylamine	0.070	291.5	8.8 [283]	−1590	−1.2
H_2O–tributylphosphinoxide	0.023	284.9	5.9 [284]	−130	−0.20
H_2O–methyldi(n-amyl)phosphinoxide	0.020	292.4	4.5 [284]	−29	−0.03
D_2O–2,4-dimethylpyridine	0.056	288.4	3.4 [279]	−660	−0.43
D_2O–2,6-dimethylpyridine	0.068	301.7	3.9 [279]	−690	−0.61
D_2O–3-methylpyridine	0.073	311.7	2.0 [285]	−250	−0.30

Table 4.4: Main values defining the thermodynamic similarity of the solutions with lower critical dissolution temperature: (1) H_2O–2,4-dimethylpyridine; (2) H_2O–2,6-dimethylpyridine; (3) H_2O–triethylamine; (4) H_2O–tributylphosphinoxide; (5) H_2O–methyldi(n-amyl)phosphinoxide; (6) D_2O–2,4-dimethylpyridine; (7) D_2O–2,6-dimethylpyridine; (8) D_2O–3-methylpyridine.

	T_D, K	p_D MPa	C	$\left(\frac{d\widetilde{p}_c}{d\widetilde{T}_c}\right)_{p=0}$	$\widetilde{h}_c^E$	$100\widetilde{v}_c^E$	$\frac{\widetilde{h}_c^E}{\widetilde{v}_c^E}$	$\widetilde{T}_0^{-1}$	Ref.
(1)	297.1	40	16.5	92.0	−0.203	−0.55	36.9	213	[287]
(2)	312.3	110	8.0	29.9	−0.267	−1.95	13.2	53.9	[246]
(3)	327.0	540	4.0	3.1	−0.585	−23.8	2.46	9.2	[253]
(4)	298.0	248	5.8	8.8	−0.073	−2.20	3.27	22.6	[286]
(5)	300.5	120	9.5	8.2	−0.012	−0.15	8.00	37.1	[286]
(6)	289.6	37	16.0	117	−0.274	−0.66	41.5	241	[253]
(7)	306.2	100	8.1	34.6	−0.271	−2.40	11.3	68.0	[253]
(8)	315.3	50	12.0	33.1	−0.095	−0.57	16.7	92.7	[288]

dissolution temperature line has the form shown in Fig. 4.13, and the double critical point is located within a pressure range from several tens to several hundreds of MPa and is available for experimental observation. There is literature containing the required information about phase separation of these systems at atmospheric pressure and about the behavior of their critical curves.

Table 4.3 contains the properties characterizing phase separation at atmospheric pressure for the studied solutions. Also in the table are references to the works containing the experimental (T, x)-data employed for the calculation of the binodal amplitude B at atmospheric pressure. The parameter B has been calculated using the least squares method in the temperature range $0 < (T - T_c)/T_c < 0.01$. The excess volume and enthalpy in the critical point for the water–triethylamine solution were taken respectively from Refs. [280] and [281]. For other solutions the values of excess volume and enthalpy were measured by the authors of the present book.

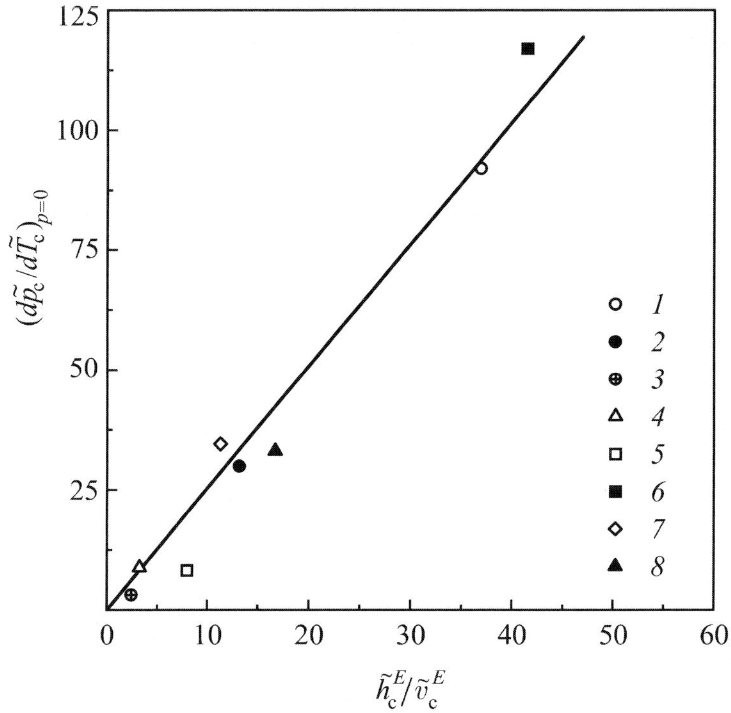

Figure 4.17: Correlation between the quantities $\left(d\widetilde{p}_c/d\widetilde{T}_c\right)_{p=0}$ and $(\widetilde{h}_c^E/\widetilde{v}_c^E)$ for aqueous solutions of some organic compounds: (1) H_2O–2,4-dimethylpyridine, (2) H_2O–2,6-dimethylpyridine, (3) H_2O–triethylamine, (4) H_2O–tributylphosphinoxide, (5) H_2O–methyldi(n-amyl)phosphinoxide, (6) D_2O–2,4-dimethylpyridine, (7) D_2O–2,6-dimethylpyridine, (8) D_2O–3-methylpyridine.

Table 4.4 contains the characteristic values defining the thermodynamic similarity of the investigated solutions. The references to the literature containing the data for the critical point lines of the solutions are given in the last column of the table. The parameter C in Eq. (4.57), and the values T_D and p_D for the water–triethylamine solution were calculated by experimental data, because the critical point line measured up to 440 MPa [253] does not reveal a double critical point in this pressure range. For the water–tributylphosphinoxide solution the dependence of the decomposition temperature on pressure is determined in Ref. [286] for the concentration which is close to the critical one, $x_2 = 0.036$, so in the calculations the values $v^E = -0.022$ cm^3/mol and $h^E = -180$ J/mol are used as measured for this concentration.

Figures 4.17–4.19 show the established correlations between dimensionless thermodynamic complexes for the investigated systems [279]. The correlations shown in Figs. 4.17 and 4.18 are described by the simple formulas

$$\left(\frac{d\widetilde{p}_c}{d\widetilde{T}_c}\right)_{p=0} = 2.587\frac{\widetilde{h}_c^E}{\widetilde{v}_c^E} , \tag{4.65}$$

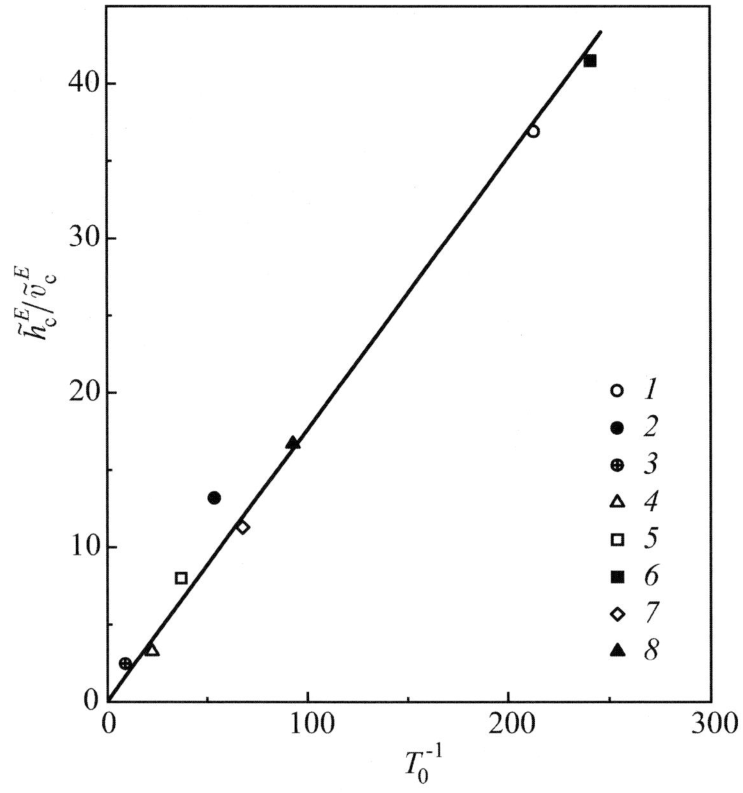

Figure 4.18: Correlations between the parameters $(\widetilde{h}_c^E/\widetilde{v}_c^E)$ and $(\widetilde{T}_0^{-1})$ for aqueous solutions of some organic compounds. The notations are the same as in Fig. 4.17.

$$\frac{\widetilde{h}_c^E}{\widetilde{v}_c^E} = 0.175\widetilde{T}_0^{-1} \, . \tag{4.66}$$

The dependence presented in Fig. 4.19 is expressed by the simple equation

$$\ln\left(\frac{C}{Bx_{2c}}\right) = 0.590 \ln \left|\widetilde{v}_c^E\right|^{-1} + 1.219 \tag{4.67}$$

or

$$\frac{C}{Bx_{2c}} = 3.384 \left|\widetilde{v}_c^E\right|^{-0.590} \, . \tag{4.68}$$

Using the obtained correlations and Eq. (4.62) and having the data for the decomposition of the mixture at atmospheric pressure $\left((T,x) - \text{data},\ v_c^E,\ h_c^E\right)$ one can calculate the values T_D, p_D, C, $(dp/dT_c)_{p=0}$ and construct the line of critical points of the solutions. The set of equations Eqs. (4.63), (4.65)–(4.67) is reduced then to the following equation

$$\frac{1}{Bx_{2c}}\left(\frac{T_D}{T_D - T_0}\right)^{-0.090} = 1.109 \left|\frac{RT_D}{h_c^E}\right|^{0.590} , \tag{4.69}$$

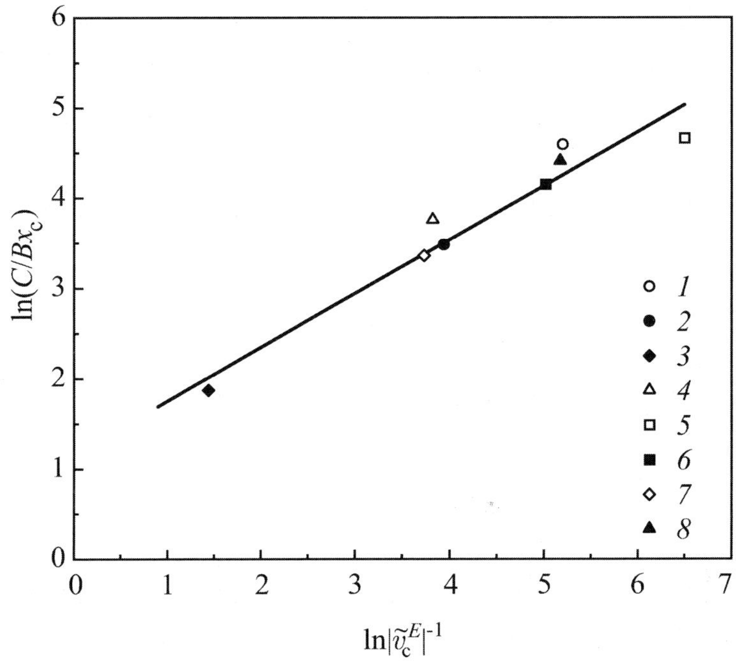

Figure 4.19: Correlations between the parameters $\ln(C/Bx_c)$ and $\ln|\tilde{v}_c^E|^{-1}$ for aqueous solutions of organic compounds. The notations are the same as in Fig. 4.17.

and solving it one obtains the unknown temperature T_D. The pressure p_D can be determined from Eq. (4.66), and for the calculation of the parameter C we get a simple formula

$$C = 1.091 \left(\frac{T_D}{T_D - T_0} \right)^{0.5} . \tag{4.70}$$

Then the value of the parameter D is determined from Eq. (4.59) at $p_c = 0$.

We choose the water–2,6-dimethylpyridine solution for the verification of the proposed algorithm of calculation of the line of critical points (Eqs. (4.63), (4.65)–(4.67)) by the data related to atmospheric pressure. The calculated parameters of Eq. (4.59) for this system have the following values: $T_D = 311.5$ K, $p_D = 117.3$ MPa, $C = 9.39$, and $D = 6.67$. Table 4.5 contains both experimental and calculated data for the critical point line of the solution in the pressure range 0.1–230 MPa. There is a good agreement of these data at pressures inside and below the region of the double critical point and some difference at high pressures. The existing divergencies are caused by three empiric correlations obtained relying on the analysis of experimental data of different precision for a group of solutions. In particular, the graphical data for the critical point lines from the literature [253, 286, 287] were used.

The established similarity relations for the solutions with lower critical dissolution temperature point to the effectiveness of the proposed approach, which is based on the use of the double critical point as a reference state. Such an approach can be useful for studying systems with a double critical point of a different nature like ferroelectrics or liquid crystals.

Table 4.5: Comparison of calculated and experimental [246] values of the critical dissolution temperature of the water–2,6-dimethylpyridine solution in the pressure range 0.1–230 MPa.

p, MPa	$T_{c(exp)}$, K	$T_{c(calc)}$, K
0.1	307.3	307.31
5.0	307.60	307.69
24.5	309.30	308.98
49.0	310.81	310.18
73.5	311.80	310.98
98.0	312.19	311.40
108.0	312.21	311.48
127.0	312.08	311.48
147.0	311.63	311.28
176.5	310.44	310.66
202.0	308.86	309.84
226.5	306.82	308.81

4.7 Concluding Remarks

In contrast to the analysis performed for one-component systems, the study of the behavior of phase-separating solutions has some fragmentary character. The fragmentary character of the present chapter is due to the large qualitative variety of phase equilibriums in solutions. Some characteristics related to two-component systems were chosen here. Also the derivation of the generalized Clausius–Clapeyron equation for this case (known as the van der Waals equation for solutions) is given in a convenient form which was used in the present investigation employing the ideal solution model.

The application of the thermodynamic model of ideal solutions allowed us to obtain correlations for the distribution of components between liquid–vapor and crystal–liquid coexisting phases. The advantage of such an approximation is demonstrated and the transition to a more precise solution model using thermodynamic similarity concepts is outlined. In the case of two liquid phases, which are different by composition, there are critical phenomena that make these solutions closer to the liquid–vapor equilibrium in single-component systems. At the same time, the isopleths $x_1(p, T) = \text{constant}$ can be represented by the Simon equation, which is characteristic for the melting lines of elementary substances. These properties allow us to establish new correlations for the thermodynamic parameters related to the coexistence of the phases. For more complex systems with a double critical point the thermodynamic similarity of isopleths can be revealed using this point as a reference one. Such an approach allows us to propose the algorithm allowing us to calculate the unknown critical point line of a solution by the data for decomposition behavior at atmospheric pressure.

A Appendices

A.1 List of Symbols

A	Filippov's thermodynamic similarity parameter
B	amplitude of the binodal curve
c_p	isobaric specific heat capacity, J/(kg K)
c_v	isochoric specific heat capacity, J/(kg K)
E	activation energy, J
G_*	Gibbs's number
h	Planck's constant, $\mathrm{h} = 6.63 \cdot 10^{-34}$ J·s
h	specific enthalpy, J/kg
Δh	change of specific enthalpy at phase transformation, J/kg
h^E	molar excess enthalpy of a solution, J/mol
$\widetilde{h}^E = h^E/RT$	reduced excess enthalpy of a solution with upper critical dissolution temperature
$\widetilde{h}^E = h^E/RT_D$	reduced excess enthalpy of the solution with lower critical dissolution point
h^M	molar heat of mixing of a solution, J/mol
J	steady-state nucleation rate, $1/(\mathrm{s\,m^{-3}})$
k_B	Boltzmann constant, $k_B = 1.380662 \cdot 10^{-23}\,\mathrm{J/K}$
N_A	Avogadro's number, $N_A = 6.022045 \cdot 10^{23}/\mathrm{mol}$
N_1	number of molecules in a unit volume of a metastable phase, $1/\mathrm{cm}^3$
N_i	number of molecules of the component i in a binary system
p	pressure, Pa
p_i	internal pressure, Pa
p_*	characteristic pressure of the condensed phase in the Simon equation at $T = 0$ K, Pa
$\widetilde{p}_{SL} = (p + p_*)/p_*$	reduced melting pressure
$\widetilde{p}_{LV} = p_{LV}/p_c$	reduced pressure of the saturated vapor
$\widetilde{p}_{LL} = (p + p_*)/p_*$	reduced separation pressure of a solution with upper critical dissolution point
$\widetilde{p}_c = (p_c - p_D)/p_D$	reduced critical pressure for a mixture with a lower critical dissolution point
R	universal gas constant, $R = 8.314$ J/(mol K)
r	radius of a spherical nucleus in a metastable phase, nm
r_*	radius of the critical nucleus, nm

Crystal–Liquid–Gas Phase Transitions and Thermodynamic Similarity. Vladimir P. Skripov and Mars Z. Faizullin

ISBN: 3-527-40576-3

s	specific entropy, $\mathrm{J/(kg \cdot K)}$
s^E	molar excess entropy of the solution, J/(mol K)
Δs	change of the specific entropy at the phase transition, $\mathrm{J/(kg \cdot K)}$
Δs_h	variation of the entropy of the liquid at homophase variations of the volume, $\mathrm{J/(kg \cdot K)}$
T	absolute temperature, K
T_0	temperature of the phase transition at atmospheric pressure, K
T_g	glass transition temperature, K
$\widetilde{T}_{SL} = T/T_0$	reduced melting temperature
$\widetilde{T}_{LV} = T/T_c$	reduced temperature on the liquid–vapor equilibrium coexistence line
$\widetilde{T}_{LL} = T/T_0$	reduced separation temperature for systems with an upper critical dissolution point
$\widetilde{T}_c = (T_D - T_c)/T_D$	reduced critical dissolution temperature for systems with a lower critical dissolution point
ΔT	supercooling, K
u	specific internal energy, J/kg
Δu	change of the specific internal energy at the phase transformation, J/kg
v	specific volume, m^3/kg
Δv	change of the specific volume at the phase transformation, $\mathrm{m^3/kg}$
v^E	molar excess volume of the solution, $\mathrm{cm^3}$/mol
$\widetilde{v}^E = v^E (p + p_*)/RT_0$	reduced excess volume of the solution with upper critical dissolution point
$\widetilde{v}^E = v^E p_D/RT_D$	reduced excess volume of the solution with lower critical dissolution point
W_*	work of critical nucleus formation, J
x	molar concentration
α	phase separation coefficient in a binary system
α_p	thermal expansion coefficient, $1/\mathrm{K}$
β	coefficient of substance distribution between the different phases in a binary system
β_T	isothermal compressibility, $1/\mathrm{Pa}$
γ_i	activity coefficient of the component i in a binary system
η	dynamic viscosity, $\mathrm{Pa \cdot s}$
μ	chemical potential, J/kg
ρ	density, $\mathrm{kg/m^3}$
$\widetilde{\rho}_L = \rho_L/\rho_c$	reduced density of the liquid on the saturation line
$\widetilde{\rho}_V = \rho_V/\rho_c$	reduced density of the saturated vapor
σ	surface tension, $\mathrm{J/m^2}$
τ	life-time of the metastable phase, s

A.2 Superscripts and Subscripts

The superscripts specify

$'$	the low-temperature phase
$''$	the high-temperature phase;

The subscripts specify

c	the critical point
D	the double critical point of a solution
L	the liquid phase
S	the crystal phase
V	the vapor phase
LL	the liquid–liquid equilibrium
LV	the liquid–vapor equilibrium
SL	the crystal–liquid equilibrium
1	the first component of the solution
2	the second component of the solution

References

[1] V. P. Skripov, *Metastable Liquids* (Nauka, Moscow, 1972; Wiley, New York, 1974).

[2] L. D. Landau and E. M. Lifshitz, *Statistical Physics* (Nauka, Moscow, 1976; Academy of Sciences Publishing House, Berlin, 1987).

[3] V. K. Semenchenko, *Selected Topics in Theoretical Physics* (Prosveshenie, Moscow, 1966).

[4] J. W. Gibbs, *Thermodynamics, Statistical Mechanics* (Nauka, Moscow, 1982); *Collected Works, Vol. 1: Thermodynamics* (Longmans & Green, New York/London/Toronto, 1928).

[5] M. Volmer, *Kinetik der Phasenbildung* (Steinkopff, Dresden-Leipzig, 1939).

[6] V. P. Skripov and V. P. Koverda, *Spontaneous Crystallization of Supercooled Liquids* (Nauka, Moscow, 1984).

[7] Ya. B. Zeldovich, Zhurn. Exper. Teor. Fiz. **12**, 525 (1942).

[8] Ya. I. Frenkel, *Kinetic Theory of Liquids* (Nauka, Leningrad, 1975; Clarendon, Oxford, 1946).

[9] A. G. Stoletov, *Selected Papers* (Gostekhteorizdat, Moscow-Leningrad, 1950).

[10] D. I. Mendeleev, Chemical Journal (edited by N. Sokolov and A. Engelhardt) **3**, 81 (1860).

[11] T. Andrews, Phil. Trans. R. Soc. London **159**, 575 (1869); T. Andrews, *On the Continuity of the Gaseous and Liquid States of Matter* (Gostekhteorizdat, Moscow-Leningrad, 1933).

[12] D. I. Mendeleev, Poggendorfs Annalen der Physik **141**, 618 (1870).

[13] J. D. van der Waals, *Über die Kontinuität des gasförmigen und flüssigen Zustandes* (Roth, Leipzig, 1881).

[14] V. P. Skripov, E. N. Sinitsyn, P. A. Pavlov, G. V. Ermakov, G. N. Muratov, N. V. Bulanov, and V. G. Baidakov, *Thermophysical Properties of Liquids in the Metastable State* (Gordon and Breach, London, 1988).

[15] R. Balescu, *Equilibrium and Non-Equilibrium Statistical Mechanics*, volume 1 (Wiley, New York/London/Sydney/Toronto, 1975; Mir, Moscow, 1978).

[16] M. P. Vukalovich and I. I. Novikov, *Equations of State of Real Gases* (Gosenergoizdat, Moscow-Leningrad, 1948).

[17] Ya. I. Frenkel, *Introduction into the Theory of Metals* (Nauka, Leningrad, 1972).

[18] V. P. Skripov, V. G. Baidakov, S. P. Protsenko, and V. V. Malzev, Teplofiz. Vysokikh Temperatur **11**, 682 (1973).

Crystal–Liquid–Gas Phase Transitions and Thermodynamic Similarity. Vladimir P. Skripov and Mars Z. Faizullin

ISBN: 3-527-40576-3

[19] M. P. Vukalovich and I. I. Novikov, *Thermodynamics* (Mashinostroenie, Moscow, 1972).

[20] E. E. Shpilrain and P. M. Kesselmann, *Basis of the Theory of Thermo-Physical Properties* (Energiya, Moscow, 1977).

[21] L. P. Filippov, *Similarity of the Properties of Matter* (Moscow State University Publishers, Moscow, 1978).

[22] K. S. Pitzer, D. Z. Lippmann, R. F. Curl, Ch. M. Huggins, and D. E. Petersen, J. Amer. Chem. Soc. **77**, 3433 (1955).

[23] L. Riedel, Chem. Ing. Technik **26**, 83 (1954).

[24] H. E. Stanley, *Introduction to Phase Transitions and Critical Phenomena* (Oxford University Press, Oxford, 1971).

[25] M. A. Anisimov, *Critical Phenomena in Liquids and Liquid Crystals* (Nauka, Moscow, 1987).

[26] V. P. Skripov (the results of this section were presented for the first time at the 8th Research Workshop *Nucleation Theory and Applications* in Dubna, Russia, October 10, 2004).

[27] D. Yu. Ivanov, L. A. Makarevich, and O. N. Sokolov, Pisma v Zhurn. Eksper. Teor. Fiz. **20**, 272 (1974).

[28] D. Yu. Ivanov and L. A. Makarevich, Doklady Akademii Nauk USSR **220**, 1103 (1975).

[29] N. Kurzeya, Th. Tielkes, and W. Wagner, Int. J. Thermophys. **20**, 531 (1999).

[30] D. Yu. Ivanov, *Critical Behavior of Non-Ideal Systems* (Fizmatlit., Moscow, 2003).

[31] E. A. Lynton, *Superconductivity* (Methuen, London,1969).

[32] R. Reid, D. Prausnitz, and T. Sherwood, *Properties of Gases and Liquids* (McGraw-Hill, New York, 1977; Khimiya, Leningrad, 1982).

[33] L. P. Filippov, *Methods of Computation and Prediction of the Properties of Matter* (Moscow State University Publishers, Moscow, 1988).

[34] E. Fermi, *Thermodynamics* (Prentice Hall, New York, 1937).

[35] V. P. Skripov and M. Z. Faizullin, Doklady Akademii Nauk **360**, 471 (1998).

[36] V. P. Skripov and M. Z. Faizullin, Teplofizika Vysokikh Temperatur **37**, 152 (1999).

[37] V. A. Rabinovich, A. A. Wassermann, V. I. Nedostup, and L. S. Veksler, *Thermophysical Properties of Neon, Argon, Krypton, and Xenon*, (Izd. Standartov, Moscow, 1976; translation from Russian in: Monograph of the National Standard Reference Data Service of the USSR (Hemisphere, New York, 1988)).

[38] V. V. Sychev, A. A. Wassermann, A. D. Kozlov, G. A. Spiridonov, and V. A. Tsymarnyi, *Thermodynamic Properties of Oxygen* (Izdatelstvo Standartov, Moscow, 1981).

[39] V. V. Sychev, A. A. Wassermann, A. D. Kozlov, G. A. Spiridonov, and V. A. Tsymarnyi, *Thermodynamic Properties of Nitrogen* (Izdatelstvo Standartov, Moscow, 1977).

[40] N. B. Vargaftik, *Tables on Thermophysical Properties of Liquids and Gases* (Fizmatizdat, Moscow, 1963; Wiley, New York, 1975).

[41] K. Schefler, J. Straub, and U. Grigull, *Wasserdampftafeln* (Springer, Berlin/Heidelberg/New York, 1981).

[42] V. V. Sychev, A. A. Wassermann, V. A. Zagorutchenko, A. D. Kozlov, G. A. Spiridonov, and V. A. Tsymarnyi, *Thermodynamic Properties of Methane* (Izdatelstvo Standartov, Moscow, 1979).

[43] I. K. Kikoin and A. P. Senchenko, Fiz. Metallov i Metallovedenie **24**, 843 (1967).

[44] E. E. Shpilrain, K. A. Yakimovich, E. E. Totzkyi, D. L. Timrot, and V. A. Fomin, *Thermophysical Properties of Alkali Metals* (Izdatelstvo Standartov, Moscow, 1970).

[45] E. E. Shpilrain, K. A. Yakimovich, S. N. Skovorodko, and A. G. Mozgovoy, *Overview on the Thermophysical Properties of Matter No. 6(44): The Density and Thermal Expansion of Liquid Alkali Metals* (IVTAN, Moscow, 1983).

[46] V. P. Skripov and M. Z. Faizullin, Int. J. Thermophys. **21**, 1213 (2000).

[47] V. G. Baidakov, *The Interface of Simple and Quantum Liquids* (Nauka, Ekaterinburg, 1994).

[48] R. Z. Magaril, Zhurn. Fiz. Khimii **29**, 1301 (1955).

[49] L. P. Filippov, Zhurn. Fiz. Khimii **54**, 2979 (1980).

[50] G. N. Muratov, Zhurn. Fiz. Khimii **56**, 1562 (1982).

[51] M. Z. Faizullin, Fluid Phase Equilibrium **211**, 75 (2003).

[52] G. N. Muratov, E. N. Sinitsyn, and V. P. Skripov, *The Capillary Constant and the Surface Tension of Freons F-11, F-21, F-113*. In: *Atomic and Molecular Physics* (Ural Polytechnical Institute Publishing House, Sverdlovsk, 1972, p. 85–90).

[53] E. N. Sinitsyn, L. A. Mikhalevich, O. P. Yankovskaya, I. F. Gulezkaya, V. B. Ivakin, G. N. Muratov, and G. V. Ermakov, *Thermophysical Properties of Liquid Fluoroorganic Compounds* (Nauka, Ekaterinburg, 1995).

[54] V. V. Sychev, A. A. Wassermann, V. A. Zagorutchenko, A. D. Kozlov, G. A. Spiridonov, and V. A. Tsymarnyi, *Thermodynamic Properties of Ethane* (Izdatelstvo Standartov, Moscow, 1982).

[55] V. V. Sychev, A. A. Wassermann, A. D. Kozlov, and V. A. Tsymarnyi, *Thermodynamic Properties of Propane* (Izdatelstvo Standartov, Moscow, 1989).

[56] S. M. Stishov, Uspekhi Fiz. Nauk **96**, 467 (1968).

[57] N. Kowai and Y. Inokuti, Japanese J. Appl. Phys. **9**, 31 (1970).

[58] D. A. Kirzhnitz, Uspekhi Fiz. Nauk **104**, 489 (1971).

[59] A. R. Ubbelohde, *Melting and Crystal Structure* (Clarendon Press, Oxford, 1965; Mir, Moscow, 1969).

[60] H. D. Luedemann and G. C. Kennedy, J. Geophys. Res. **73**, 2795 (1968).

[61] J. E. Lennard-Jones and A. F. Devonshire, Proc. R. Soc. A **147**, 464 (1939).

[62] L. D. Landau, Zhurn. Eksper. Teor. Fiz. **7**, 19 (1937).

[63] P. W. Bridgman, *The Physics of High Pressure* (G. Bell, London, 1931; ONTI, Moscow-Leningrad, 1935).

[64] P. W. Bridgman, *Recent Works in the Field of High Pressure* (Rev. Mod. Phys. **18**, 1 (1946); Izd. Inostrannoi Literatury, Moscow, 1948).

[65] A. Michels, B. Blaisse, and J. Hoogschagen, Physica **9**, 565 (1942).

[66] M. Z. Faizullin, *Measurements of the Melting Temperature of Fluoroorganic Substances at Pressures up to 250 MPa*. In: *Phase Transformations in Metastable Systems* (Ural Branch of the Academy of Sciences, Sverdlovsk, 1983).

[67] M. Z. Faizullin, Zhurn. Fiz. Khimii **59**, 245 (1985).
[68] F. E. Simon and G. Glatzel, Zs. Anorg. Chem. **178**, 309 (1929).
[69] F. P. Bandy, J. Chem. Phys. **38**, 618 (1963).
[70] L. F. Vereshchagin and N. S. Fateeva, Zhurn. Exper. Teor. Fiz. **55**, 1145 (1968).
[71] E. I. Asinovskij, A. V. Kirillin, and A. V. Kostanovskij, Teplofizika Vysokikh Temperatur **35**, 716 (1997).
[72] V. D. Urlin, Zhurn. Eksper. Teor. Fiz. **49**, 485 (1965).
[73] R. Boehler, M. Ross, and D. B. Boercker, Phys. Rev. B **53**, 556 (1996).
[74] E. Yu. Tonkov, *Phase Diagram of the Elements at High Pressures* (Nauka, Moscow, 1979).
[75] V. M. Glasov, V. B. Lazarev, and V. V. Zharov, *Phase Diagrams of Simple Substances* (Nauka, Moscow, 1980).
[76] L. Merrill, J. Phys. Chem. Ref. Data **6**, 1205 (1977).
[77] E. Yu. Tonkov, *Phase Diagram of Compounds at High Pressures* (Nauka, Moscow, 1983).
[78] S. E. Babb, Rev. Mod. Phys. **35**, 400 (1963).
[79] A. R. Regel and V. M. Glazov, *The Periodic Law and the Physical Properties of Electronic Alloys* (Nauka, Moscow, 1978).
[80] E. A. Guggenheim, J. Chem. Phys. **13**, 253 (1945).
[81] A. P. Kazragis, A. A. Raudeljunene, and Ya. A. Scheschtokene, Zhurn. Fiz. Khimii **50**, 2368 (1976).
[82] H. Sawamure, Trans. Japn. Inst. Metals **13**, 225 (1972).
[83] S. M. Stishov, Uspekhi Fiz. Nauk **114**, 3 (1974).
[84] S. M. Stishov, I. N. Makarenko, and A. M. Nikolaenko, Fiz. Tverdogo Tela **18**, 2863 (1976).
[85] V. I. Zubov, Zhurn. Fiz. Khimii **55**, 2171 (1981).
[86] F. A. Lindemann, Phys. Zs. **11**, 609 (1910).
[87] S. M. Stishov, Uspekhi Fiz. Nauk **154**, 93 (1988).
[88] M. Lasocka, Phys. Lett. A **51**, 137 (1975).
[89] R. M. Cotterill and J. L. Tallon, Faraday Disc. Chem. Soc. **69**, 241 (1980).
[90] J. L. Tallon, Phys. Lett. A **76**, 139 (1980).
[91] J. U. Madsen and R. M. Cotterill, Phys. Lett. A **83**, 219 (1981).
[92] J. L. Tallon, Phys. Lett. A **87**, 361 (1982).
[93] J. L. Tallon, Solid State Commun. **42**, 243 (1982).
[94] J. L. Tallon, Phys. Rev. B **29**, 4153 (1984).
[95] A. Rubcic and J. Baturic-Rubcic, Phys. Lett. A **72**, 27 (1979).
[96] A. Rubcic and J. Baturic-Rubcic, Fizika **12**, 253 (1980).
[97] J. Baturic-Rubcic and A. Rubcic, Fizika **12**, 259 (1980).
[98] S. J. Henderson and R. J. Speedy, J. Phys. Chem. **91**, 3069 (1978).
[99] V. P. Skripov, Inzhen. Fiz. Zhurnal **59**, 431 (1990).
[100] V. P. Skripov and M. Z. Faizullin, Teplofiz. Vysokikh Temperatur **37**, 814 (1999).

[101] V. P. Skripov and M. Z. Faizullin, Zhurn. Fiz. Khimii **59**, 598 (1985).

[102] Ch. Tegler, R. Span, and W. Wagner, J. Phys. Chem. Ref. Data **28**, 779 (1999).

[103] I. N. Makarenko, A. M. Nikolaenko, V. A. Ivanov and S. M. Stishov, Zhur. Eksper. Teor. Fiz. **69**, 1723 (1975).

[104] M. A. Pokrasin, V. V. Roshchupkin, L. R. Fokin, and N. E. Khandamirova, *Interpolation Equations and the Tables of Saturated Vapor Pressure of Sodium, Potassium, Rubidium and Cesium in the Temperature Range from the Triple Point up to the Critical Point. The Equation for the Saturation Pressure of Lithium up to 2500 K.* In: *Thermophysical Properties of Substances and Materials* (Izdatelstvo Standartov, Moscow, 1983, p. 33–55).

[105] V. M. Cheng, W. B. Daniels, and R. K. Crawford, Phys. Rev. B **11**, 3972 (1975).

[106] V. V. Altunin, *Thermophysical Properties of Carbon Dioxide* (Izdatelstvo Standartov, Moscow, 1975).

[107] I. F. Golubev, V. P. Kiyashova, I. I. Perelshtein, and E. B. Parushin, *Thermophysical Properties of Ammonia* (Izdatelstvo Standartov, Moscow, 1978).

[108] B. A. Younglove and J. L. Ely, J. Phys. Chem. Ref. Data **16**, 577 (1987).

[109] L. E. Reeves, G. J. Scott, and S. E. Babb, J. Chem. Phys. **40**, 3662 (1964).

[110] J. H. Dymond, J. Robertson, and J. D. Isdale, J. Chem. Thermodynam. **14**, 51 (1982).

[111] A. Würflinger, Ber. Bunsenges. Phys. Chem. **79**, 1195 (1975).

[112] A. M. Mamedov, T. S. Akhundov, and F. G. Abdullaev, *Thermophysical Properties of Liquid Benzene*. In: *Thermophysical Properties of Substances and Materials* (Izdatelstvo Standartov, Moscow, 1973, p. 71–84).

[113] M. K. Zhokhovskij and V. S. Bogdanov, Zhurn. Fiz. Khimii **39**, 2520 (1965).

[114] M. A. McCool and L. A. Woolf, J. Chem. Soc. Faraday Trans. I **10**, 1971 (1972).

[115] G. C. Straty and R. Tsumura, J. Chem. Phys. **64**, 895 (1976).

[116] E. S. Domalski and E. D. Hearing, J. Phys. Chem. Ref. Data **25**, 1 (1996).

[117] Yu. L. Vasiliev, Izvestiya Vuzov: Neftch i Gaz **10**, 57 (1984).

[118] B. A. Grigoriev, Yu. L. Rastorguyev, A. A. Gerasimov, D. S. Kurumov, and S. A. Plotnikov, *Thermodynamic Properties of Normal Hexane* (Izdatelstvo Standartov, Moscow, 1990).

[119] Yu. L. Vasiliev, Izvestiya Vuzov: Neftch i Gaz **5**, 49 (1985).

[120] J. H. Dymond, J. Robertson, and J. D. Isdale, Int. J. Thermophys. **2**, 133 (1981).

[121] A. Würflinger and G. M. Schneider, Ber. Bunsenges. Phys. Chem. **77**, 121 (1973).

[122] R. Landau and A. Würflinger, Ber. Bunsenges. Phys. Chem. **84**, 895 (1980).

[123] A. K. Doolittle, J. Chem. Eng. Data **9**, 275 (1964).

[124] J. L. Daridon, H. Carrier, and B. Lagourette, Int. J. Thermophys. **23**, 697 (2002).

[125] A. A. Schärer, C. J. Busso, A. E. Smith, and L. B. Skinner, J. Amer. Chem. Soc. **77**, 2017 (1955).

[126] W. G. Cutler, R. H. McMickle, W. Webb, and R. W. Schiessler, J. Chem. Phys. **29**, 727 (1958).

[127] S. Dutour, J. L. Daridon, and B. Lagourette, Int. J. Thermophys. **21**, 173 (2000).

[128] S. Dutour, B. Lagourette, and J. L. Daridon, J. Chem. Thermodynam. **33**, 765 (2001).

[129] S. Dutour, B. Lagourette, and J. L. Daridon, J. Chem. Thermodynam. **34**, 475 (2002).

[130] B. Koppitz and A. Würflinger, Colloid Polymer Sci. **252**, 999 (1974).

[131] I. S. Grigoriev and E. Z. Meilikhov (Eds.), *Physical Quantities: Encyclopedia* (Energoatomizdat, Moscow, 1991)

[132] P. M. Kesselmann and V. P. Onishchenko, *Thermodynamic Properties of Liquid Alkali Metals in the One-Phase Region*. In: *Thermophysical Properties of Substances and Materials*, vol. 11 (Izdatelstvo Standartov, Moscow, 1977, pp. 145–166).

[133] V. E. Zinovyev, *Thermophysical Properties of Metals at High Temperatures* (Metallurgiya Publishers, Moscow, 1989).

[134] I. I. Novikov, Yu. S. Trelin, and T. A. Tsyganova, Teplofizika Vysokikh Temperatur **10**, 1114 (1972).

[135] I. I. Novikov, Yu. S. Trelin, and T. A. Tsyganova, Teplofizika Vysokikh Temperatur **8**, 450 (1970).

[136] M. P. Vukalovich, A. I. Ivanov, L. R. Fokin, and A. T. Yakovlev, *Thermophysical Properties of Mercury* (Izdatelstvo Standartov, Moscow, 1971).

[137] P. W. Mirwald and G. C. Kennedy, J. Geophys. Res. **84**, 6750 (1979).

[138] F. Simon, Trans. Faraday Soc. **33**, 65 (1937).

[139] V. P. Skripov, Zhurn. Fiz. Khimii **56**, 546 (1982).

[140] V. P. Skripov and M. Z. Faizullin, *The Behavior of the Thermodynamic Stability and the Viscosity Along the Melting Curve*. In: *Thermophysical Properties of Metastable Systems* (Ural Branch of the Russian Academy of Sciences, Sverdlovsk, 1984, pp. 8–16).

[141] V. P. Skripov and M. Z. Faizullin, High Temp.-High Press. **18**, 1 (1986).

[142] V. G. Baidakov, V. P. Skripov, and A. M. Kaverin, Zhurn. Eksper. Teor. Fiz. **67**, 676 (1974).

[143] P. H. Lahr and W. G. Eversole, J. Chem. Eng. Data **72**, 42 (1962).

[144] M. M. Martynyuk, Zhurn. Fiz. Khimii **57**, 810 (1983).

[145] A. D. Kirshenbaum, J. A. Cahill, P. J. McGonigal, and A. V. Grosse, J. Inorg. Nucl. Chem. **24**, 1287 (1962).

[146] J. K. Krause and C. A. Swenson, Cryogenics **16**, 413 (1976).

[147] G. C. Straty and R. Pryds, Phys. Lett. A **31**, 301 (1970).

[148] S. E. Babb, J. Chem. Phys. **50**, 5270 (1969).

[149] V. G. Baidakov, A. E. Galashev, and V. P. Skripov, Fiz. Tverdogo Tela **22**, 2681 (1980).

[150] V. I. Zubov and V. B. Magalinskij, *On the Thermodynamic Stability of the Crystalline Phase*. In: *Thermophysical Properties of Metastable Systems* (Ural Branch of the Russian Academy of Sciences, Sverdlovsk, 1984, pp. 42–48).

[151] M. N. Krivoguz and G. E. Norman, Doklady Akademii Nauk **379**, 177 (2001).

[152] C. A. Swenson, Phys. Rev. **99**, 423 (1955).

[153] R. L. Beecroft and C. A. Swenson, J. Phys. Chem. Solids **18**, 329 (1961).

[154] V. N. Zharkov and V. A. Kalinin, *Equations of State of Solids at High Pressures and Temperatures* (Nauka, Moscow, 1968).

[155] N. H. Macmillan, *The Ideal Strength of Solids*. In: *Atomistic Approach to Fracture* (Plenum Press, New York, 1983, pp. 95–164).
[156] V. P. Skripov, Teplofiz. Vysokikh Temperatur **19**, 85 (1981).
[157] V. E. Lyusternik, Teplofiz. Vysokikh Temperatur **28**, 686 (1990).
[158] V. V. Brazhkin and A. G. Lyapin, Uspekhi Fiz. Nauk **170**, 535 (2000).
[159] H. J. Parkhurst and J. Jonas, J. Chem. Phys. **63**, 2705 (1975).
[160] S. A. Ulybin and W. I. Makarushkin, *The Viscosity of Carbon Dioxide at 220–1300 K and Pressure up to 300 MPa*. In: *Proceedings of the 7th Symposium on Thermophysical Properties* (Gaithersburg, New York, 1977, pp. 678–683).
[161] N. J. Trappeniers, P. S. van der Gulik, and H. van den Hooff, Chem. Phys. Lett. **70**, 438 (1980).
[162] A. I. Batchinskij, *Selected Works* (Izdatelstvo Akademii Nauk of the USSR, Moscow, 1960).
[163] H. Vogel, Phys. Z. **22**, 645 (1921).
[164] G. Tammann, *Der Glaszustand* (Leopold Voss Verlag, Leipzig, 1933; ONTI, Moscow-Leningrad, 1935).
[165] O. V. Mazurin, *Vitrification* (Nauka, Leningrad, 1986).
[166] A. K. Doolittle, J. Appl. Phys. **22**, 1471 (1951).
[167] D. L. Hogenboom, W. Webb, and J. A. Dixon, J. Chem. Phys. **46**, 2586 (1967).
[168] L. S. Serdjuk, *Thermodynamic Properties of Normal Hydrogen at Temperatures up to 1500 K and Pressures up to 5000 bar*. In: *Thermophysical Properties of Substances and Materials* (Izdatelstvo Standartov, Moscow, 1973, pp. 3–5).
[169] I. I. Novikov, Yu. S. Trelin, and T. A. Tsyganova, Teplofiz. Vysokikh Temperatur **7**, 1220 (1969).
[170] *Properties of the Condensed Phases of Hydrogen and Oxygen. Encyclopedia* (Naukova Dumka, Kiev, 1984).
[171] R. K. Crawford and W. B. Daniels, J. Chem. Phys. **50**, 3171 (1969).
[172] V. P. Skripov and M. Z. Faizullin, Doklady Akademii Nauk **378**, 620 (2001).
[173] H. Suga and S. Seki, J. Non-Cryst. Solids **16**, 171 (1974).
[174] C. A. Angell, J. M. Sare and E. J. Sare, J. Phys. Chem. **82**, 2622 (1978).
[175] T. Atake and C. A. Angell, J. Phys. Chem. **83**, 3218 (1979).
[176] V. P. Skripov and M. Z. Faizullin, *Behavior of the Volume and Entropy Jumps Along the Melting Curve and Its Low-Temperature Extension*. In: *Thermophysical Properties of Metastable Systems* (Ural Branch of the Russian Academy of Sciences, Sverdlovsk, 1983, pp. 18–23).
[177] V. A. Ivanov, I. N. Makarenko, A. M. Nikolaenko, and S. M. Stishov, Phys. Lett.A **47**, 75 (1974).
[178] I. N. Makarenko, A. M. Nikolaenko, and S. M. Stishov, Zhurn. Eksper. Teor. Fiz. **74**, 2175 (1978).
[179] V. P. Skripov and M. Z. Faizullin, Doklady Akademii Nauk **363**, 618 (1998).
[180] P. W. Bridgman, Phys. Rev. **3**, 153 (1914).
[181] D. V. Dokhov, Metally **4**, 28 (1999).

[182] V. G. Baidakov, *Superheating of Cryogenic Liquids* (Ural Branch of the Russian Academy of Science, Ekaterinburg, 1995).

[183] D. G. Thomas and L. A. Staveley, J. Chem. Soc. **12**, 4569 (1952).

[184] H. Hollomon and D. Turnbull, *Formation of Nuclei in Phase Transitions*. In: *Uspekhi Fiziki Metallov*, vol. 1 (BHTIL Metallurgii, Moscow, 1956, pp. 304–367); *Nucleation*. In: *Progress in Metal Physics* (Pergamon Press, London, 1953).

[185] V. P. Skripov and M. Z. Faizullin, Doklady Akademii Nauk **372**, 749 (2000).

[186] V. P. Skripov and M. Z. Faizullin, Zhurn. Fiz. Khimii **75**, 670 (2001).

[187] V. P. Skripov, V. P. Koverda, and G. T. Butorin, Kristallografiya **15**, 1219 (1970).

[188] G. S. Zhdanov, Fiz. Tverdogo Tela **18**, 1415 (1976).

[189] V. P. Koverda, V. N. Skokov, and V. P. Skripov, Kristallografiya **27**, 358 (1982).

[190] E. Yu. Tonkov, *Phase Transitions of Compounds at High Pressure*, vols. 1 and 2 (Metallurgiya, Moscow, 1988).

[191] M. E. Cavalery, T. G. Plymate, and J. H. Stout, J. Phys. Chem. Solids **49**, 945 (1988).

[192] P. W. Mirwald and G. C. Kennedy, J. Phys. Chem. Solids **37**, 795 (1976).

[193] V. P. Skripov and M. Z. Faizullin, *On the Possibility of Computation of the Melting Curve Employing Thermodynamic Similarity Methods*. In: *Thermophysical Properties of Substances and Materials*,vol.3 (Izdatelstvo Standartov, Moscow, 1986, p. 49–60).

[194] H. N. V. Temperley, J. S. Rowlinson, and G. S. Rushbrooke, *Physics of Simple Liquids: Statistical Theory* (North-Holland, Amsterdam, 1968; Mir, Moscow, 1971).

[195] V. P. Skripov, Fiz. Metallov Metallovedenie **57**, 421 (1984).

[196] V. A. Ivanov, I. N. Makarenko, and S. M. Stishov, Pisma v Zhurn. Eksper. Teor. Fiz. **12**, 12 (1970).

[197] I. N. Makarenko, V. A. Ivanov, and S. M. Stishov, Pisma v Zhurn. Eksper. Teor. Fiz. **18**, 320 (1973).

[198] V. P. Skripov and M. Z. Faizullin, Rasplavy **2**, 3 (1988).

[199] J. U. Madsen and R. M. Cotteril, Phys. Scripta **24**, 959 (1981).

[200] J. D. Bernal, Nature (London) **183**, 141 (1959).

[201] V. P. Skripov and A. E. Galashev, Uspekhi Khimii **52**, 177 (1983).

[202] A. E. Galashev and V. P. Skripov, Zhurn. Strukturnoi Khimii **25**, 77 (1984).

[203] V. P. Skripov and A. E. Galashev, Kristallografiya **27**, 961 (1982).

[204] P. W. Bridgman, Phys. Rev. **6**, 1, 94 (1915).

[205] H. Schinke and F. Sauerwald, Z. Phys. Chem. **216**, 26 (1961).

[206] V. P. Skripov, M. Z. Faizullin, and A. V. Shteinert, Zhurn. Fiz. Khimii **61**, 344 (1987).

[207] J. Jackson, Phys. Earth Planet. Int. **14**, 8694 (1977).

[208] S. P. Clark, J. Chem. Phys. **31**, 1526 (1959).

[209] C. W. F. T. Pistorius, J. Chem. Phys. **45**, 3513 (1966).

[210] C. W. F. T. Pistorius, J. Phys. Chem. Solids **26**, 1543 (1965).

[211] C. W. F. T. Pistorius, J. Chem. Phys. **43**, 1557 (1965).

[212] P. Pawlow, Z. Phys. Chem. **65**, 1, 545 (1908).

[213] H. Reiss and I. B. Wilson, J. Colloid. Sci. **3**, 551 (1948).

[214] K. J. Hanszen, Z. Physik **157**, 523 (1960).

[215] Ph. Buffat and J. P. Borel, Phys. Rev. A **13**, 2287 (1976).

[216] P. R. Couchman and W. A. Jesser, Nature (London) **269**, 481 (1977).

[217] V. P. Koverda, V. N. Skokov, and V. P. Skripov, phys. stat. sol. (a) **74**, 343 (1982).

[218] V. P. Skripov, V. P. Koverda, and V. N. Skokov, phys. stat. sol. (a) **66**, 109 (1981).

[219] M. Takagi, J. Phys. Soc. Japan **9**, 359 (1954).

[220] L. S. Palatnik and Yu. F. Komnik, Fiz. Metallov Metallovedenie **9**, 374 (1960).

[221] B. K. Wainshtein, *Structural Electronography* (Academy of Sciences of the USSR, Moscow, 1956).

[222] V. P. Koverda, V. N. Skokov, and V. P. Skripov, Fiz. Metallov Metallovedenie **51**, 1240 (1981).

[223] V. P. Skripov, Zhurn. Fiz. Khimii **72**, 2102 (1998).

[224] B. N. Srivastava and R. P. Rastogi, Proc. Natl. Inst. Sci. India **19**, 613 (1953).

[225] J. D. van der Waals and Ph. Kohnstamm, *Lehrbuch der Thermostatik, Erster Teil* (Johann Ambrosius Barth Verlag, Leipzig, 1927); *A Course in Thermostatics* (ONTI, Moscow, 1936).

[226] I. Prigogine and R. Defay, *Chemical Thermodynamics* (Longmans Green, London/New York/ Toronto, 1954; Nauka, Novosibirsk, 1966).

[227] D. ter Haar and G. Vergeland, *Elements of Thermodynamics* (Addison-Wesley, Reading, Massachusetts, 1966; Mir, Moscow, 1968).

[228] V. P. Skripov, Teplofiz. Vysokikh Temperatur **36**, 898 (1998).

[229] V. V. Kafarov (Ed.), *Encyclopedia on Liquid–Vapor Phase Equilibrium* (Goskhimizdat, Leningrad, 1957).

[230] V. P. Skripov and E. N. Dubrovina, Zhurn. Fiz. Khimii **72**, 2106 (1998).

[231] V. P. Skripov and E. N. Dubrovina, Doklady Akademii Nauk **357**, 213 (1997).

[232] V. P. Skripov, Zhurn. Obshchei Khimii **70**, 1772 (2000).

[233] A. I. Brodskij, *Chemistry of Isotopes* (Academy of Sciences of the USSR, Moscow, 1952).

[234] V. P. Skripov and L. V. Povyshev, Zhurn. Fiz. Khimii **36**, 325 (1962).

[235] A. Carli, S. Di Cave, and E. Sebastiani, Chem. Eng. Sci. **27**, 993 (1972).

[236] E. N. Dubrovina and V. P. Skripov, *Determination of One of the Equilibrium Curves of Binary Systems from the Other Employing the Model of Perfect Solutions*. In: *Metastable States and Phase Transitions*, vol. 6 (Ural Branch of the Russian Academy of Sciences, Ekaterinburg, 2003, pp. 42–48).

[237] P. I. Bystrov, D. N. Kagan, G. A. Kretchetova, and E. E. Shpilrain, *Liquid-Metal Heat Carriers of Thermal Tubes and Energetic Equipment* (Nauka, Moscow, 1988).

[238] E. N. Dubrovina and V. P. Skripov, *Thermodynamic Similarity and the Equilibrium Phase Diagram Crystal–Liquid for Metallic Systems K–Cs, Rb–Cs, K–Rb, Cu–Ni, Cu–Au*. In: *Metastable States and Phase Transitions*, vol. 3 (Ural Branch of the Russian Academy of Sciences, Ekaterinburg, 1999, pp. 91–97).

[239] I. F. Holscher, G. M. Schneider, and J. B. Ott, *Liquid–Liquid Phase Equilibria of Binary Mixtures of Methanol with Hexane, Nonane and Decane at Pressures up to 150 MPa*. In: *Studies in Modern Thermodynamics*, vol. 6 (Elsevier, Amsterdam, 1986, pp. 153–169).

[240] J. Timmermans, J. Chim. Physique **20**, 491 (1923).

[241] V. F. Alekseev, Gornyi Zhurnal **6**, 385 (1885).

[242] M. Z. Faizullin and V. A. Shteinert, *Optical High-Pressure Camera for the Investigation of Phase Separation of Liquid Solutions at Pressures up to 200 MPa*. In: *Phase Transitions and High-Energy Processes* (Ural Branch of the Russian Academy of Sciences, Ekaterinburg, 1988, pp. 112–115).

[243] M. Z. Faizullin and V. P. Skripov, Rasplavy **2**, 46 (1988).

[244] V. P. Skripov and M. Z. Faizullin, Zhurn. Fiz. Khimii **62**, 3247 (1988).

[245] V. P. Skripov and M. Z. Faizullin, J. Chem. Thermodynam. **21**, 687 (1989).

[246] M. Z. Faizullin and V. P. Skripov, J. Chem. Thermodynam. **23**, 561 (1991).

[247] M. Z. Faizullin and V. P. Skripov, J. Chem. Thermodynam. **26**, 1167 (1994).

[248] M. Z. Faizullin and V. P. Skripov, *The Phase Separation of Water and Methydi(n-amyl)phosphynoxide at Pressures up to 230 MPa*. In: *Process Technology Proceedings, 12. High Pressure Chemical Engineering* (Elsevier, Amsterdam, 1996, pp. 481–485).

[249] K. Roth, G. M. Schneider, and E. U. Frank, Ber. Bunsenges. Phys. Chem. B **70**, 5 (1966).

[250] G. M. Schneider, *High Pressure Thermodynamics and Phase Equilibria of Fluid Mixtures: Survey and Recent Results*. In: *High Pressure Science and Technology*, Proc. 7th Int. AIRAPT Conf. Le Creosot, vol. 2 (Oxford, 1980, pp. 685–691).

[251] G. M. Schneider and R. Lecture, J. Chem. Thermodynam. **23**, 301 (1991).

[252] R. Grzanna and G. M. Schneider, Z. Phys. Chem. B **193**, 41 (1996).

[253] G. M. Schneider, Ber. Bunsenges. Phys. Chem. B **70**, 497 (1966).

[254] L. C. van den Bergh, J. A. Schouton, and N. J. Trappeniers, Physica A **141**, 524 (1987).

[255] L. C. van den Bergh and J. A. Schouton, Chem. Phys. Lett. **145**, 471 (1988).

[256] H. K. Schurmann and R. D. Parks, Phys. Rev. Lett. **26**, 367 (1971).

[257] R. B. Pettit and W. J. Camp, Phys. Rev. Lett. **32**, 369 (1974).

[258] V. M. Glazov and S. G. Kim, Zhurn. Fiz. Khimii **61**, 2171 (1987).

[259] V. M. Glazov, S. G. Kim, and G. K. Mamberterzina, Zhurn. Fiz. Khimii **66**, 2912 (1992).

[260] D. V. Sokolskij, V. F. Kiselev, and R. Sh. Nigmetova, Doklady Akademii Nauk **295**, 655 (1987).

[261] R. A. Khairulin and S. V. Stankus, Zhurn. Fiz. Khimii **70**, 1230 (1996).

[262] G. I. Pozharskaya, N. L. Kasapova, V. P. Skripov, and Yu. D. Kolpakov, Zhurn. Fiz. Khimii **59**, 1822 (1985).

[263] A. E. Nesterov and Yu. S. Lipatov, *Phase State of Solutions and Mixtures of Polymers. Encyclopedia* (Kiev, Naukova Dumka, 1987).

[264] H. Endo, H. Hoshino, K. Tamura, and M. Mushiage, Solid State Commun. **32**, 1243 (1979).

[265] K. Tamura, H. Hoshino, and H. Endo, Ber. Bunsenges. Phys. Chem. **84**, 235 (1980).

[266] D. B. Myers, R. A. Smyth, J. Katz, and R. L. Scott, J. Phys. Chem. **70**, 3341 (1966).

[267] V. P. Skripov and G. I. Pozharskaya, *On the Thermodynamic Similarity of Phase Separating Solutions*. In: *Thermophysics of Metastable Liquids in Application to the Processes of Boiling and Crystallization* (Ural Branch of the Russian Academy of Sciences, Sverdlovsk, 1987, pp. 144–146).

[268] R. Paas and G. M. Schneider, J. Chem. Thermodynam. **11**, 267 (1979).

[269] Yu. P. Blagoi and A. V. Savina, *Excess Solution Volumes of Liquefied Gases of the System* $CH_4 - CF_4$, $CH_4 - C_3H_6$, $CH_4 - Kr$. In: *Physics of the Condensed State*, vol. 5 (Collected Works of the Low-Temperature Physico-Technical Institute of the Ukrainian Academy of Sciences, Kharkov, 1969, pp. 15–22).

[270] V. P. Belousov, A. G. Morachevskiy, and M. Yu. Panova, *Thermal Properties of Non-Electrolyte Solutions. Encyclopedia* (Khimiya Publishers, Leningrad, 1981).

[271] R. W. Kiser, G. D. Johnson, and M. D. Shetlar, J. Chem. Eng. Data **6**, 338 (1961).

[272] H. Wolff and K. Bernstorff, Ber. Bunsenges. Phys. Chem. **62**, 1093 (1958).

[273] O. Azocar and J. Edwards, Monatshefte Chem. **102**, 1866 (1971).

[274] D. R. Thompson and O. K. Rice, J. Amer. Chem. Soc. **86**, 3547 (1964).

[275] A. Deerenberg, J. A. Schouten, and N. J. Trappeniers, Physica A **103**, 183 (1980).

[276] H. W. Bruckman, A. C. Michels, and N. J. Trappeniers, Physica A **139**, 175 (1986).

[277] H. M. J. Boots and A. C. Michels, Physica A **103**, 316 (1980).

[278] M. A. Anisimov, A. V. Voronel, and E. E. Gorodetzkij, Zhurn. Eksper. Teor. Fiz. **60**, 1117 (1971).

[279] M. Z. Faizullin and V. P. Skripov, Z. Phys. Chem. B **173**, 53 (1991).

[280] F. Kohler, Monatshefte Chem. B **82**, 913 (1951).

[281] J. L. Copp and D. H. Evert, Disc. Faraday Soc. **15**, 174 (1953).

[282] R. J. L. Andon and J. D. Cox, J. Chem. Soc. **12**, 4601 (1952).

[283] F. Kohler and O. K. Rice, J. Chem. Phys. **26**, 1614 (1957).

[284] A. V. Nikolaev and I. I. Yakovlev, *Clathrate Formation and Physico-Chemical Analysis of Extraction Systems* (Nauka, Novosibirsk, 1975).

[285] J. D. Cox, J. Chem. Soc. **11**, 4606 (1952).

[286] A. V. Nikolaev, P. N. Kuznetsov, D. S. Mirinskiy, Yu. A. Dyadin, I. I. Yakovlev, and N. S. Patrin, Doklady Akademii Nauk USSR **223**, 101 (1975).

[287] G. Schneider, Z. Phys. Chem., Neue Folge **39**, 187 (1963).

[288] G. Schneider, Z. Phys. Chem., Neue Folge **37**, 333 (1963).

Index